PLA Modernisation and Likely Force Structure: 2025

PLA Modernisation and Likely Force Structure: 2025

by

Colonel Nagender S P Bisht

United Service Institution of India

New Delhi

Vij Books India Pvt Ltd

New Delhi (India)

Published by

Vij Books India Pvt Ltd
(Publishers, Distributors & Importers)
2/19, Ansari Road
Delhi – 110 002
Phones: 91-11-43596460, 91-11-47340674
Fax: 91-11-47340674
e-mail: vijbooks@rediffmail.com

ISBN (HB)   : 978-93-84464-50-9
ISBN (PB)   : 978-93-84464-49-3
ebook ISBN   : 978-93-84464-51-6

Paperback edition published in 2016

Contents

Preface   ix

Acknowledgements   xi

Abbreviations   xiii

Introduction   1

Chapters

1.   PLA: Historical Perspective, its War Fighting Doctrines
        and RMA with Chinese Characteristics   5

2.   PLA Army (PLAA) Modernisation   37

3.   PLA Air Force (PLAAF) Modernisation   75

4.   PLA Navy (PLAN) Modernisation   100

5.   PLA Second Artillery Force (PLASAF) Modernisation   119

6.   Dragon Fire and Indian Response   133

Conclusion   157

Appendices

Appendix 1   159

        Some identifications of formations and sub units in
        Chengdu and Lanzhou MAC

Appendix 2   162

        Deployment of PAPF in Chengdu and Lanzhou MAC
        Reserve formations/ units in the Chengdu and the Lanzhou MR

Appendix 3   166

        Present Deployment of the PLAAF

Appendix 4   174

Modernization Trends PLAAF 2009 Onwards

Appendix 5   186

Transformation of PLAN from 2007 to 2013

Appendix 6   197

An Analysis of the Military Balance and Equipment
Holdings

Annexure 1   212

Revolution in Military Affairs with Chinese Characteristics

Bibliography   225

Index   227

List of Tables and Maps

Tables

Table 1.1:	Timelines Evolution of PLA War fighting Doctrines	18
Table 2.2:	Analysis PLAA Armoured Formations	42
Table 2.3:	Analysis Transformation of Group Armies	43
Table 2.4:	Approximate Strength of PLAA Formations and Units	46
Table 2.5:	Trends of Transformation MAC	47
Table 3.1:	Details of Chinese Military Satellites	82
Table 3.2:	Organization of 15 AB Corps	90
Table 3.3:	Summary of Deployment PLAAF	92
Table 4.1:	North Sea Fleet Details	109
Table 4.2:	East Sea Fleet Details	110
Table 4.3:	South Sea Fleet Details	111
Table 4.4:	Future Trends PLAN	112
Table 5.1:	Command and Control Structure PLA SAF	121
Table 5.2:	Details PLA SAF Brigades	122
Table 5.3:	PLA SAF Missile Systems and Launchers	124
Table 5.4:	Analysis of PLA SAF Brigades and Missile Launchers	126
Table 5.5:	Trends PLA SAF Missile Holdings	127
Table 6.1:	Analysis of Military Balance in a China – Pakistan Collusive Scenario	141
Table 6.2:	Group Armies – Primary and Secondary Missions	148

Table 6.3:   Likely force levels that may be applied against India in any contingency (in 2014)   149

Maps

Map 3.1:   Present Deployment of PLAAF   91

Map 4.1:   Present Deployment of PLAN   108

Foreword

The significance of China's emergence as a global power in early 21st century is hard to overemphasize. The prospect of a non-democratic, Asian nation surpassing America's economic supremacy in the near future is both daunting and far-reaching in its implications at the global geopolitical level. Although averse to towing the US line, China has so far shied away from playing an assertive role on the international stage. However, the tide is turning and the secretive dragon is gradually flexing its military muscle to protect and enhance its geo-strategic interests.

As the Asian giant transitions from an export-led to a consumption-driven economy, Deng Xiaoping's 'peaceful rise' mantra is gradually giving way to a more insular and militaristic outlook. Striving to control havens of raw materials and to secure trade routes, China is raking up old unsettled feuds with neighbours and betraying greater hegemonic ambitions. This is evident in its disputes over islands in the South China Sea and East China Sea, frequent violation of the border with India, recurrent altercations with Japan's new government and fractious exchanges with the US over trade and currency issues.

This rising assertiveness is complemented by broad-based efforts to modernize and expand the People's Liberation Army (PLA). Over the last decade, China's officially disclosed military budget has grown at an annual rate of about 10 percent. The PLA is also increasing its presence in various regions around the world and enhancing its ability to execute long-range mobility operations.

This book provides an in-depth study and analysis of the ongoing, comprehensive Modernisation program of the PLA and extrapolates the trend up to 2025. It intensively explores the entire gamut of PLA's Modernisation, covering the PLA Army (PLAA), PLA Air Force (PLAAF), PLA Navy (PLAN) and the PLA Second Artillery Force (PLASAF). After providing historical perspective on the origins of the PLA, the book analyzes subsequent shifts in its war doctrines over the years. Following an insightful analysis of the Modernisation of the four services, the book details the manifestation and impact of this process. Every chapter concludes with details of the appreciated force structure by the year 2025.

The book is a valuable resource in terms of its exhaustiveness and analytical insight on this important subject of contemporary military studies. It will prove to be an indispensable compendium for every student of the PLA's Modernisation drive and force structure in the years ahead.

November 2014

Lt Gen P.K. Singh,
PVSM, AVSM (Retd)
Director, United Service Institution of India

Preface

中国 (Zhongguo) is the most common name for China in simplified Chinese and is written using two characters. The first character 中 (zhong) means 'central' or 'middle' and the second character 国 (guo) means 'state' or 'country'. This name has been there for almost last two thousand years and finds mention in the Classic of History or the 'Shangshu' as the name for centre of civilization or an explanation for the 'Tianxia' concept. The basic concept being that China is the 'centre to the world'; the Emperor is the 'son of heaven' and rules over 'all under the heaven' and all others are barbarians, who have to be dealt with. This fact each Chinese has been learning for last two thousand years and is now embedded in the Chinese thought process from a common man to their leadership.

The Bronze Age saw the first civilizations develop in the fertile river valleys of Euphrates and Tigris in Mesopotamia, Nile in Egypt, Indus in Indian subcontinent and the Huang he or the Yellow river in China. Through the passage of time, today all these civilizations and their people, less the Chinese, have undergone huge and irreversible changes. China, isolated from the remaining world, owing to geography has retained a large part of this culture and thinking that has developed through the passage of time and today she has the most homogenous population with similar culture, traditions, eating habits and thinking that still has large influences from her past.

Through the last two thousand years the idea of China and the thought of the middle kingdom have developed in a manner that the glimpses of this we see in the present Chinese thinking. The stamp released depicting the China dream (zhong guo meng) has the Great Wall in the background so was there when President Xi addressed the nation from the Presidential office in Zhongnanhai on the New Year in 2014. Today China feels that the century of humiliation is over, China is rising and to continue rising she has to be strong militarily. It is in the background of this thought that the pace of Chinese military Modernisation needs to be studied.

This book is a study of the development and Modernisation of the Peoples Liberation Army (PLA) from 1921 onwards, focusing on the trends of development during the revolution and the civil war, formation of the Peoples Republic of China and extrapolating it into future up to 2025. The world has changed a lot in last 25 years and this pace of change has been the fastest in the known history less the drastic changes due to natural disasters. If any country has kept pace or even at times out paced these changes, that is China and the same is true for the Chinese military. India an old civilization with an equally developed civilizational thought, an emerging economic and military power with a favourable demographic profile needs to watch the military developments in China very carefully, continuously and in detail. This becomes even more critical in light of China-Pakistan relations and India- China border dispute, in addition to a host of other reasons including economy and trade. India also has to develop her capabilities to match the Chinese rise and ensure a safe and secure future in the Asian century.

In last twenty years one has seen the rise of China and the PLA. Weather it is the development of infrastructure Tibet or the stupendous economic rise, one has been in awe. A perception that China will be a major adversary in times ahead has taken shape in minds of many in India including me. Is it true? Is the PLA progressing so fast and India will not be able to match up? Is the PLA Modernisation without hurdles? What will be the PLA capabilities by 2025? And many such questions arose in the mind and then I thought of researching this field and try finding answers to my questions. This is my first book and is a result of my experiences of last 22 years of military service and about two years of research.

Acknowledgements

To research and study an emerging and assertive China indeed was a great honor and for which I am grateful to the Indian Army for the two years study leave. I place on record my gratitude to the United Service Institution of India (USI of India) for accepting my research proposal and providing a congenial, stimulating and academically conducive environment for my research. My sincere and heartfelt gratitude to Lieutenant General P K Singh, PVSM, AVSM (retired), Director, USI of India and Major General Y K Gera (retired), Consultant and Head (Research) at the Centre for Strategic Studies and Simulation for encouraging and providing necessary support and help to complete my research.

One person without whose guidance and constant course correction this study would not have been possible is Major General B K Sharma, AVSM, SM**(retired), my guide for this study. He always gave me a constructive advice and suggestions, discussed the study regularly in detail and was instrumental in shaping the final outcome of the book. Sincere and heartfelt thanks to him.

My gratitude to Professor Nirmala Joshi and Doctor Roshan Khaniejo fur their timely comments, suggestions and recommendations which were a great help. Both the ladies are experts in their own fields and are highly competent and experienced in research. It was my singular honor to be associated with them.

I sincerely thank Brigadier Saif Ul Islam Khan, a fellow Research Scholar, with whom I shared not only the research room but also a large number of thoughts and issues. These were of great help in not only completing this book but also in clearing a number of thoughts and concepts. My sincere gratitude to Lieutenant General Vinod Bhatia, PVSM, AVSM, SM (retired), Major Genral S V Thapliyal, SM (retired) and Professor Srikant Kondapalli of the Jawaharlal Nehru University for their invaluable suggestions and recommendations.

I benefited enormously with my interactions with Brigadier A S Cheema, VSM (retired) and other scholars of CS3, USI of India. Their perceptive observations and suggestions really helped me in shaping the final outcome of my research.

Lastly I must accept that nothing would have been possible without the support and constant help from my wife Major Monisha Dhami (retired), my daughter Nayanika, who edited a large part of this book, corrected the grammar and spellings and son Aarush. They were the support without which it may not have been possible for me to bring this endeavour to its logical end.

USI of India   Nagender S P Bisht
New Delhi
October 2014

Abbreviations

AAA	Anti-Aircraft Artillery
AAR	Air to air refuelling
AEW	Airborne early Warning
AEW&C	Airborne Early Warning and Control
AMS	Academy of Military Sciences
APC	Armoured Personnel Carrier
ATGM	Anti-Tank Guided Missile
AWACS	Airborne Warning and Control System
BDS BeiDou	("Big Dipper" constellation) Navigation Satellite System
C3I	Command, Control, Communications and Intelligence
C^4I^2SR	Command, Control, Communications, Computers, Intelligence, Information, Surveillance and Reconnaissance
CAS	Close Air Support
CC	Combined Corps
CMC	Central Military Commission
COSTIND	Commission of Science, Technology and Industry for National Defence
CP	Command Post
CPC	Communist Party of China
CPV	Chinese People's Volunteers

CRS	Congressional Research Service (US)
ELINT	Electronic Intelligence
DF	Dong Feng ("East Wind")
DZ	Drop Zone (for airborne operations)
ECM	Electronic Countermeasure
GA	Group Army
GAD	General Armaments Department
GLD	General Logistics Department
GLONASS	Global Navigation Satellite System (Russia)
GPD	General Political Department
GPS	Global Positioning Satellite system
GSD	General Staff Department
HJ Hongjian	("Red Arrow")
HN Hongying	(variously "Red Tassel" or "Red Cherry")
HQ Hongqi	("Red Flag")
ICBM	Intercontinental Range Ballistic Missile
IFV	Infantry Fighting Vehicle
Il	Ilysuhin
IRBM	Intermediate Range Ballistic Missile
IT	Information Technology (sometime followed by the word "application")
IW	Information Warfare
KMT	Kuomintang (Nationalist Party)
LSM	Medium Landing Ship
LST	Tank Landing Ship
LZ	Landing Zone (for airmobile operations)

MD	Military District (also known as Provincial Military Command)
MND	Ministry of National Defence
MPS	Ministry of Public Security
MR	Military Region
MAC	Military Area Command
MSD	Military Sub district
MSS	Ministry of State Security
MTEP	Military Training and Evaluation Program (also called the Outline for Military Training and Testing)
MUCD	Military Unit Code Designator
NCO	Non-commissioned Officer
NCW	Net Centric Warfare
NDMC	National Defence Mobilization Committee
NDU	National Defence University
NEFA	North East Frontier Agency
NORINCO	China North Industries Corporation
NUDT	National Defence Science and Technology University (also known as the National University of Defence Technology)
PAFD	People's Armed Forces Departments
PAPF	People's Armed Police Force
PGM	Precision-Guided Munitions
PKO	Peacekeeping Operation
PLA	People's Liberation Army
PLAA	People's Liberation Army Army
PLAAF	People's Liberation Army Air Force

PLAN	People's Liberation Army Navy
PLA SAF	People's Liberation Army Second Artillery Force
PRC	People's Republic of China
QW Qianwei	("Vanguard")
RMA	Revolution in Military Affairs
RRU/F	Rapid Reaction (or Response) Unit /Force
SAM	Surface-to-Air Missile
SAR	Search and Rescue
SIGINT	Signal Intelligence
SOF	Special Operations Forces
SP	Self-Propelled
SRBM	Short-Range Ballistic Missile
SSM	Surface to Surface Missile
Su	Sukhoi
TO&E	Table of Organization and Equipment
Tu	Tupolev
UAV	Unmanned Aerial Vehicle
UCAV	Unmanned Combat Aerial Vehicle
UI	Unidentified
UN	United Nations
US DoD	US Department of Defence

Introduction

"What is essential in war is victory, not prolonged operations."

- Sun Tzu, The Art of War

"Let China sleep, for when she awakes, she will shake the world."

-Napoleon

A superpower is a state with a dominant position in the international system which has the ability to influence events and its own interests and project power on a worldwide scale to protect those interests. A superpower is traditionally considered to be a step higher than a great power. Alice Lyman Miller (Professor of National Security Affairs at the Naval Postgraduate School), defines a superpower as "a country that has the capacity to project dominating power and influence anywhere in the world, and sometimes, in more than one region of the globe at a time, and so may plausibly attain the status of global hegemony"[1].

China has a long military tradition, dating back to the earliest days of recorded history. The martial exploits of kings and emperors, loyal generals and peasant rebels, strategists and theorists are well known in Chinese culture and folk tradition. Study of Chinese military history reveals that throughout the centuries, two trends have influenced the role of the military in China, one during the times of peace and the other in times of unrest and anarchy. In times of peace and stable regimes the military was firmly under the control of the civilian authorities. The military had the capability and capacity to deal with both domestic disorders and foreign invasions, yet it did not threaten the existing political system. In times of disorder, however, new military leaders and organizations rose to challenge the old system, resulting in the militarization of political life. When one of these leaders became strong enough, he established a new political order ruling over all of China. After consolidating power, the new ruler or his successors subordinated the military to civilian control once again. In the past 150 years, a third factor entered the Chinese military tradition: the introduction of modern military technology and organization to strengthen military capabilities against domestic and foreign enemies.

Since the beginning of the 20th century, all three tendencies have been discernible in the role of the military in national life. These factors have been particularly apparent in the role of the PLA in the rise to power of the CPC, in the military's role in the politics of the PRC, and in the efforts of Chinese leaders to modernize the armed forces. Post formation of the PRC, the PLA, which developed from a peasant guerrilla force to a conventional military organization capable of achieving long sought national liberation from foreign colonial powers and the Japanese occupation, in the 1950s pursued a vigorous military Modernisation program with Soviet assistance. Political involvement in the Great Leap Forward (1958–60) and the Cultural Revolution (1966–76) delayed these efforts until the late 1970s, when the PLA embarked on the next military Modernisation program, which had three major focuses. First, military Modernisation required both the strengthening of party control over the military and the continued disengagement of the armed forces from politics. These steps were necessary to ensure that a politically reliable yet professionally competent military would concentrate on the task of military reform. Secondly, defence Modernisation attempted to achieve improved combat effectiveness through organizational, doctrinal, training, educational, and personnel reforms (including recruitment, promotion, and demobilization). These reforms emphasized the development of combat capabilities in waging combined arms warfare. Thirdly, military Modernisation was aimed at the transformation of the defence establishment into a system capable of independently sustaining modern military forces. This transformation necessitated the reorganization and closer integration of civilian and military science and industry and also the selective use of foreign technology.

Post the breakup of relations in the late 1960s, China had considered the Soviet Union the principal threat to its security; lesser threats were posed by long-standing border disputes with Vietnam and India. China's territorial claims and economic interests made the South China Sea an area of strategic importance to China. Although China sought peaceful reunification of Taiwan with the mainland China, she did not rule out the use of force against the island if serious internal disturbances, a declaration of independence, or a threatening alliance occurred.

The scope of foreign military cooperation has evolved gradually. In the 1950s, China dealt only with communist nations and insurgencies. In the 1960s, it began to provide military assistance to Third World nations to counteract Soviet and US influence. Beginning in the late 1970s, China shifted its arms transfer policy away from military assistance in favour of commercial arms sales and began developing military ties with Western Europe and US. Chinese military contacts with foreign countries expanded rapidly with the introduction of the PLA military Modernisation program and the policy of opening up to the outside world.

In the late 1980s, People's Liberation Army forces consisted of the Army (ground forces), Air Force, Navy and Strategic Missile Force (also known as the Second Artillery Corps). The Army was divided into group armies and regional forces (Military District). Ground force equipment was largely of Soviet design and obsolete, although some weaponry had been upgraded with foreign technology. The Air Force had serious technological deficiencies despite incremental improvements of aircraft. The Navy was developing a blue-water capability and sea-based strategic forces. China possessed a small but relatively credible nuclear deterrent force with an incipient second-strike capability. Paramilitary forces consisted of the Militia, Reserve Service System, Production and Construction Corps and People's Armed Police Force.

A 'Rising China' with second largest economy will continue to fuel sustained and large increases in the PLA defence budget in coming years as increasing national-level demands for new PLA missions require increasing capability. However "Surprise" PLA Modernisation programs—such as the 2004 Yuan-class submarine development—will likely continue to emerge, particularly in those cases where service programs have not yet caught up to national requirements. As Modernisation continues and systems become more complex, the human dimensions of the PLA – education, training, personnel management—become more critical.

Areas of Study

The study primarily focuses on the study in the shifts in the PLA war fighting doctrines and the military Modernisation of the four components of the PLA: PLA Army (PLAA), PLA Air force (PLAAF), PLA Navy (PLAN) and the PLA Second Artillery Force (PLASAF). Trends of Modernisation of each of the service have been studied and researched and a likely state of that service projected by 2025. The question, why 2025? Year 2027 is the first centenary of the PLA and to showcase the achievements of PLA, they need to develop capabilities and force structure matching up to the occasion. Next important year in 2049, the centenary of the PRC and this also forms an important landmark for further PLA studies with year 2035 a midway point. Hence year 2025 is the first step for this PLA study and where PLA reaches by 2025 will have major implications on the PLA of 2035 and 2050.

The study is divided into six chapters with the projections for the PLA in 2025 forming part of the conclusion. The chapter one deals with the historical perspectives and shifts in the PLA warfighting doctrines. The historical perspective is studied primarily from the point of the development of the PLA and its leadership and the impact of these developments on the present day PLA including its geographical disposition. The shifts in the Chinese war fighting doctrines have been researched from Mao's Peoples War to the present day of Local Wars under the Conditions of Informationalization including Revolution in Military Affairs (RMA) with Chinese characteristics. In chapter two the Modernisation trends of the PLAA are discussed in detail. Also discussed in this chapter are the People's Armed Police Force (PAPF), Reserves and the Militia. The chapter three deals with the Modernisation of the PLA Air force (PLAAF) and in chapter four the Modernisation trends of the PLA Navy (PLAN) are discussed. Chapter five looks at the Modernisation of the PLA Second Artillery Force (PLASAF) in detail. The chapter six looks at an Indian response including a comparison of weapon systems and an appreciated force application by the PLA against India. The projected force structure and the capabilities form part of the conclusion, the final chapter of the book.

Endnotes

1   Alice Lyman Miller, China an Emerging Superpower, Stanford Journal of International Relations, 6(1), Winter 2005, accessed from http://web.stanford.edu/group/sjir/6.1.03_miller.html on August 24, 2014.

1

PLA: Historical Perspective, its War Fighting Doctrines and RMA with Chinese Characteristics

Our national defence will be consolidated and no imperialist will be allowed to invade our territory again. Our People's armed forces must be maintained and developed with the brave and steeled People's Liberation Army as their foundation. We will have not only a powerful army but also a powerful air force and a powerful navy.

- Mao Zedong[1]

History of the People's Liberation Army[2]

The history of the PLA begins with the founding of the CPC. The CPC has its origin in the May Fourth Movement of 1919, inspired by the Russian Revolution of 1917, was founded on July 1, 1921 in Shanghai. On August 7, 1927, the CPC Central Committee convened a meeting in Wuhan where the general principle of carrying out Agrarian Revolution and armed resistance against Kuomintang reactionaries was made.

The origin of the PLA can be traced back to the Nanchang Uprising on August 1, 1927 led by Zhu De and Zhou Enlai and the Autumn Harvest Uprising of September 7, 1927 led by Mao Zedong against the Kuomintang. Both the uprisings failed and the survivor's fled to the Jinggang Mountains along the border of Hunan and Jiangxi provinces. This was the first armed uprising by the Communists and it marked a significant change in their strategy and Mao and Zhu De went on to develop a rural-based strategy of guerrilla tactics, paving the way to the Long March of 1934

On August 1, 1927, the CPC launched an armed uprising in Nanchang, the capital of Jiangxi Province. The Nanchang Uprising led by Zhou Enlai, He Long, Ye Ting, Zhu De and Liu Bocheng gave birth to the armed forces of the CPC and marked the beginning of CPC's independent leadership over armed struggles. In spite of the failures of the two uprisings, the

CPC continued with the fight. Some important dates and timelines in the evolution of the PLA are[3];

- On December 11, 1927, led by Zhang Tailei, Ye Ting, Yun Daiying and Ye Jianying, the armed workers launched an uprising in Guangzhou to fight against the Kuomintang rightists' slaughter of members of the CPC and its supporters.

- In April 1928, led by Zhu De and Chen Yi, the Worker's and Peasant's Revolutionary Army composed of the remained troops of the Nanchang Uprising and the armed peasants from southern Hunan Province joined forces in the Jinggang Mountain with the Worker's and Peasant's Revolutionary Army established by Mao Zedong after the Autumn Harvest Uprising in the Hunan-Jiangxi border area.

- In October 1928, the Central Military Department of the CPC was set up.

- Gutian Meeting was held on December 28 and 29, 1929, the 9th Party Congress of the Fourth Army of the Chinese Workers' and Peasants' Red Army in Gutian Village, Shanghang County of Fujian Province. At the meeting, the fundamental principles of building the Chinese Workers' and Peasants' Red Army were established and the questions on how to build an army mainly composed of peasants into a new-type army of Chinese people under the absolute leadership of the CPC were answered. The Gutian Meeting is a milestone in the history of the building of the CPC and the PLA.

- In March 1930, the Central Military Department of the CPC was renamed the CMC of the CPC.

- Raising of the PLA Formations.

  - On August 23, 1930, the First Front Army of the Chinese Worker's and Peasant's Red Army was formed in Liuyang County, Hunan Province.

  - On November 7, 1931, the Fourth Front Army of the Chinese Worker's and Peasant's Red Army came into being in Huang'an County, Hubei Province.

> ➢ On November 25, 1931, the Central Revolutionary Military Commission of the Soviet Republic of China was established in Ruijin, Jiangxi Province.

- In December 1932, the Fourth Front Army of the Chinese Worker's and Peasant's Red Army marched to the Northern Sichuan Province and started to create the Sichuan-Shaanxi Border Base Area.

- On May 8, 1933, the General Headquarters of the Chinese Worker's and Peasant's Red Army was set up.

- On July 11, 1933, the Provisional Central Government of the Chinese Soviet Republic of China made a decision to set the August 1st, the date of the Nanchang Uprising, as the anniversary of the founding of the Chinese Worker's and Peasant's Red Army. Since then, the August 1st each year has become the Army Day of the PLA.

- The Long March.

  > ➢ By 1930 the Red Army was 30,000 strong which faced the Kuomintang army, which was a hundred thousand strong.

  > ➢ In spite of this imbalance, the Worker-Peasant's Red Army, using guerilla tactics, successfully resisted four "Encirclement and Suppression" Campaigns, one each in 1930 and 1933 and two in 1933, under the leadership of Mao Zedong; but the fifth time under Wang Ming, the Red Army failed as it deviated away from the guerilla tactics and tried fighting the war conventionally and thus forcing the Red Army to begin the Long March. In addition it was party internal politics also that forced the Red Army to abandon guerrilla warfare and undertake the Long March of 1934-35.

  > ➢ In October 1934, the First Front Army of the Chinese Worker's and Peasant's Red Army withdrew from the Central Base Area in Jiangxi Province and embarked on the Long March. Joining forces under the leadership of Mao and Zhu, this force became the First Worker's and Peasant's Army, or Red Army, the military arm of the Chinese Communist Party. Using the guerrilla tactics that would later make Mao Zedong internationally famous as a military strategist, the Red Army

survived several encirclement and suppression campaigns by the Kuomintang forces.

> On November 16, 1934, the 25th Army of the Chinese Worker's and Peasant's Red Army abandoned the Hubei-Henan-Anhui Border Base Area and started its Long March.

> From January 15 to 17, 1935, the Zunyi Meeting, an enlarged meeting of the Political Bureau of the CPC Central Committee, was held at Zunyi, Guizhou Province during the Long March. The Zunyi Meeting organizationally ended the domination of Wang Ming in the CPC Central Committee, established the leadership of the new Central Committee represented by Mao Zedong, establishing his leadership in the CPC and the Red Army.

> On September 16, 1935, the 25th Army of the Red Army ended its Long March by joining force with the Red Army troops in North West China at the Yongping Town of Yanchuna County of Northern Shaanxi Province. On October 19, 1935, the main force of the First Front Army of the Red Army led by the CPC Central Committee arrived at the Wuqi Town in Northern Shaanxi Province, which ended the Long March.

> The Red Army's exploits during the Long March became legendary and remain a potent symbol of the spirit and prowess of the Red Army and its successor, the PLA. During that period, Mao's political power and his strategy of guerrilla warfare gained ascendancy in the party and the Red Army.

- From December 17 to 25, 1935, the Political Bureau of the CPC Central Committee held an enlarged meeting to make the policy and strategic guide line of the Anti-Japanese National United Front.

- On February 20, 1936, the anti Japanese forces in different places in Northeast China were reorganized into the Northeast Anti-Japanese United Forces.

- On July 5, 1936, the 2nd and 6th Corps and the 32nd Army of the Red Army merged to form the Red Army's Second Front Army in the Long March.

- In July 1936, the First Route Army of the Northeast Anti-Japanese United Forces was established. In October 1936, the First, Second and the Fourth Front Armies of the Red Army jointed forces in the Shaanxi-Gansu-Ningxia Border Area.

- In mid-January 1937, the leading organs including the Central Committee and the CMC of the CPC transferred from Bao'an County to Yan'an, a city in northern Shannxi Province. In the following 10 years, Yan'an became the seat of the headquarters of the Chinese revolution.

Mao's military thought grew out of the Red Army's experiences in the late 1930s and early 1940s and formed the basis for the "People's War" concept, which became the doctrine of the Red Army and the PLA. In developing his thought, Mao drew on the works of the Chinese military strategist Sun Zi (fourth century BC) and Soviet and other theorists, as well as on the lore of peasant uprisings, such as the stories found in the classical novel Shuihu Zhuan (Water Margin) and the stories of the Taiping Rebellion. Synthesizing these influences with lessons learned from the Red Army's successes and failures, Mao created a comprehensive politico-military doctrine for waging revolutionary warfare[4].

People's War incorporated political, economic, and psychological measures with protracted military struggle against a superior foe. As a military doctrine, People's War emphasized the mobilization of the populace to support regular and guerrilla forces; the primacy of men over weapons, with superior motivation compensating for inferior technology; and the three progressive phases of protracted warfare – strategic defensive, strategic stalemate, and strategic offensive. During the first stage, enemy forces were "lured in deep" into one's own territory to overextend, disperse, and isolate them. The Red Army established base areas from which to harass the enemy, but these bases and other territory could be abandoned to preserve Red Army forces. In the second phase, superior numbers and morale were applied to wear down the enemy in a war of attrition in which guerrilla operations predominated. During the final phase, Red Army forces made the transition to regular warfare as the enemy was reduced to parity and eventually defeated[5].

In the Chinese Civil War which followed Japan's defeat after the Second World War, the Red Army, newly renamed the PLA People's Liberation Army, again used the principles of People's War in following a policy of strategic withdrawal, waging a war of attrition, and abandoning cities and communication lines to the well-armed, numerically superior Kuomintang forces. During this time the PLA converted from an essentially a guerilla force into a large army capable of large scale combined arms operations. It was during this time that the CMC established the GSD, GPD and the GLD. In 1947 the PLA launched a counter offensive during a brief strategic stalemate. By the next summer, the PLA had entered the strategic offensive stage, using conventional warfare as the Kuomintang forces went on the defensive and then collapsed rapidly on the mainland in 1949. By 1950 the PLA had seized Hainan Island and Xizang.

On January 15, 1949, the CMC decided to reorganize the regional armies of the PLA into four field armies. The forces in Northwest China were designated the First Field Army, with Peng Dehuai as commander and also serving as political commissar. The First Field Army was to comprise the 1st and 2nd Corps, and totalled 134,000 men. After 1949, the First Field Army controlled five provinces - Shaanxi, Gansu, Qinghai, Ningxia, and Xinjiang. The Second Field Army took control of PLA troops in central China, with Liu Bocheng as commander and Deng Xiaoping as commissar. It comprised the 3rd, 4th and 5th Corps, plus a special technical column and totaled 128,000 men. After 1949, the Second Field Army was stationed in southwest China and controlled five provinces - Yunnan, Giuzhou, Sichuan, Xikang (Kham region), and Tibet. The Third Field Army took control of the troops in eastern China, with Chen Yi as its commander. It comprised the 7th, 8th, 9th, and 10th Corps plus the headquarters of the special technical troops, with a total of 580,000 men. After 1949, the Third Field Army remained on China's East coast, controlling Shandong, Jiangsu, Zhejiang, Anhui, and Fujian. The PLA troops in Manchuria were designated the Fourth Field Army under Lin Bao. The army comprised the 12th, 13th, 14th, 15th Corps, Special Technical Troops, the column of Guangdong and Guangxi, and the 50th and 51st Armies.

In this deployment of the PLA forces is the origin of the MACs as they exist today. Even today, after numerous reorganizations of the MACs, there is very little change in the boundaries of the Lanzhou and Chengdu MACs, corresponding to the boundaries of the First and Second Field armies. The areas of the remaining two field armies were finally readjusted to form the five MACs.

The Peoples Republic of China (Zhōnghuá Rénmín Gònghéguó)

was founded on October 1, 1949 and the PLA became a national armed force. The PLA was a people's army created and led by the CPC and was the principal body of China's armed forces[6]. At that time it was an unwieldy, 5-million-strong peasant army and included 10,000 troops in the Air Force (founded in 1949) and 60,000 in the Navy (founded in 1950). China also claimed a militia of 5.5 million.

China's new leaders recognized the need to transform the PLA, essentially an infantry army with limited mobility, logistics, ordnance, and communications, into a modern military force[7]. The signing of the Sino-Soviet Treaty of Friendship, Alliance and Mutual Assistance in February 1950 provided the framework for defence Modernisation in the 1950s. However, the Korean War was the real watershed in armed forces Modernisation[8]. The CPVs, as the Chinese military forces in Korea were called, achieved initial success against the UN troops and managed to fight UN forces to a stalemate. Nevertheless, China's Korean War experience demonstrated PLA deficiencies and stimulated Soviet assistance in equipping and reorganizing the military. The use of unsupported infantry attacks against combined arms firepower caused serious manpower and material losses. The PLA Air force also suffered heavy losses to the superior UN forces. Finally, shortcomings in transportation and supply indicated the need to improve logistics capabilities[9].

Military Modernisation in the 1950s and 1960s

Large-scale Soviet aid in modernizing the PLA, which began in 1951, took the form of weapons and equipment, assistance in building China's defence industry, and the loan of advisers, primarily technical ones. Mostly during the Korean War years, the Soviet Union supplied infantry weapons, artillery, armour, trucks, fighter aircraft, bombers, submarines, destroyers and gunboats. Soviet advisers assisted primarily in developing a defence industry set up along Soviet organizational lines. Aircraft and ordnance factories and shipbuilding facilities were constructed and by the late 1950s were producing a wide variety of Soviet-design military equipment. Because the Soviet Union would not provide China with its most modern equipment, most of the weapons were outdated and lacked an offensive capability[10]. Both Chinese dissatisfaction with this defensive aid and the Soviet refusal to supply China with nuclear bomb blueprints partly contributed to the withdrawal of Soviet advisers in 1960[11].

In the early 1940s, China's leaders decided to reorganize the military along Soviet lines[12]. In 1954 they established the National Defence Council, Ministry of National Defence, and thirteen MRs. The PLA was reconstituted according to Soviet tables of organization and equipment. It adopted the combined arms concept of armour and artillery heavy mobile forces, which required the adoption of some Soviet strategy and tactics. PLA Modernisation according to the Soviet model also entailed creation of a professional officer corps, complete with Soviet-style uniforms, ranks, and insignia, conscription, a reserve system and new rules of discipline. The introduction of modern weaponry necessitated raising the education level of soldiers and intensifying formal military training.

The military's new emphasis on Soviet-style professionalism produced tensions between the party and the military. The party feared that it would lose political control over the military that the PLA would become alienated from a society concentrating on economic construction and that relations between officers and soldiers would deteriorate. The party reemphasized Mao's thesis of the supremacy of men over weapons and subjected the PLA to several political campaigns. The military, for its part, resented party attempts to strengthen political education, build a mass militia system under local party control, and conduct economic production activities to the detriment of military training. These tensions culminated in September 1959, when Mao Zedong replaced Minister of National Defence Peng Dehuai, the chief advocate of military Modernisation, with Lin Biao, who reversed the process of military professionalism in favour of revolutionary purity.

The ascension of Lin Biao and the complete withdrawal of Soviet assistance and advisers in 1960 marked a new stage in military development. The Soviet withdrawal disrupted the defence industry and weapons production, particularly crippling the aircraft industry. Although the military purchased some foreign technology in the 1960s, it was forced to stress self-reliance in weapons production. Lin Biao moved to restore PLA morale and discipline and to mold the PLA into a politically reliable fighting force. Lin reorganized the PLA high command, replaced the mass militia with a smaller militia under PLA control, and reformulated the Maoist doctrine of the supremacy of men over material. Lin stated that "men and material form a unity, with men as the leading factor", giving ideological justification to the reemphasis on military training. Political training, however, continued to occupy 30 to 40 percent of a soldier's time. At the same time, Lin instituted stricter party control, restored party organization at the company level, and intensified political education. In 1964 the prestige of the PLA as an exemplary, revolutionary organization was confirmed by the "Learn from the PLA" campaign. This campaign, which purported to disseminate the military's political-work experience throughout society, resulted in the introduction of military personnel into party and government organizations, a trend that increased after the Cultural Revolution began[13].

In the late 1950s and early 1960s, the PLA fought one internal and one external campaign: in Xizang against the Tibetan's and on the Sino-Indian border against India. In the first campaign, PLA forces suppressed the Tibetan's and assimilated Tibet into PRC. The Sino-Indian border war broke out in October 1962 amid the deterioration of Sino-Indian relations and mutual accusations of intrusions into disputed territory. In this brief, one month, but decisive conflict, the PLA attacked Indian positions in the NEFA (Arunachal Pradesh), penetrating to the Himalayan foothills, and in Ladakh, particularly in the Aksai Chin region. After gaining a large part if Indian territory, the PLA withdrew behind the original "line of actual control" after China announced a unilateral cease-fire. Both campaigns were limited conflicts using conventional tactics.

People's Liberation Army in the Cultural Revolution

The PLA played a complex political role during the Cultural Revolution. From 1966 to 1968, military training, conscription and demobilization, and political education virtually ceased as the PLA was ordered first to help promote the Cultural Revolution and then to reestablish order and authority[14]. Although the Cultural Revolution initially developed separately in the PLA and in the party apparatus, the PLA, under the leadership of its radical leftist leader, Lin Biao, soon became deeply involved in civilian affairs. In early 1967, the military high command was purged, and regional military forces were instructed to maintain order, establish military control, and support the "revolutionary left". Because many regional force commanders supported conservative party and government officials rather than radical mass organizations, many provincial level military leaders were purged or transferred, and Beijing ordered several main force units to take over the duties of the regional-force units. In the summer of 1967, regional military organizations came under leftist attack, Red Guard factions obtained weapons, and violence escalated. By September, the central authorities had called off the attack on the PLA, but factional rivalries between regional and main force units persisted. Violence among rival mass organizations, often backed by different PLA units, continued in the first half of 1968 and delayed the formation of revolutionary committees, which were to replace traditional government and party organizations. In July 1968, Mao abolished the Red Guards and ordered the PLA to impose revolutionary committees wherever such bodies previously had not been established[15].

Worries over military factionalism caused the leadership to curtail the Cultural Revolution and to initiate a policy of rotating military commanders and units. The Soviet invasion of Czechoslovakia, the enunciation of "the Brezhnev Doctrine", the Soviet military buildup in its Far Eastern theater, and Sino-Soviet border clashes in the spring of 1969 brought about a renewed emphasis on some of the PLA's traditional military roles. In 1969, Lin Biao launched an extensive "war preparations" campaign that included resumption of military training and military procurement, which had suffered in the first years of the Cultural Revolution, rose dramatically[16]. Military preparedness was further advanced along China's frontiers and particularly the Sino-Soviet border when the thirteen military regions were reorganized into eleven in 1970. The resulting military regions were the Shenyang, Beijing, Lanzhou, Xinjiang, Jinan, Nanjing, Fuzhou, Guangzhou (including Hainan Island), Wuhan, Chengdu, and Kunming MRs (MACs).

The PLA emerged from the more violent phase of the Cultural Revolution deeply involved in civilian politics and public administration. It had committed 2 million troops to political activities and reportedly suffered hundreds of thousands of casualties. Regional military forces were almost completely absorbed in political work. PLA units did not withdraw fully from these duties until 1974. Following the sudden death of Lin Biao in 1971, the military began to disengage from politics, and civilian control over the PLA was reasserted. Lin's supporters in the PLA were purged, leaving some high-level positions in the PLA unfilled for several years. PLA officers who had dominated provincial-level and local party and government bodies resigned from those posts in 1973 and 1974. Military region commanders were reshuffled, and some purged military leaders were rehabilitated[17]. Military representation in the national-level political organizations, following an all-time high at the Ninth National Party Congress in 1969, declined sharply at the Tenth National Party Congress in 1973.

Along with the reassertion of civilian control over the military and the return to military duties came a shift of resources away from the defence sector. Defence procurement dropped by 20 percent in 1971 and shifted from aircraft production and intercontinental ballistic missile development to the Modernisation of the ground forces and medium-range ballistic missile and intermediate-range ballistic missile development.

Military Modernisation in the 1970s

In January 1974, the PLA saw action in the South China Sea following a long-simmering dispute with the Republic of Vietnam (South Vietnam) over the Paracel Islands. South Vietnamese and PLA naval forces skirmished over three islands occupied by South Vietnamese troops, and the PLA successfully seized control of the islands in a joint amphibious operation involving 500 troops and air support[18].

By the mid-1970s, concerns among Chinese leaders about military weakness, especially vis-à-vis the Soviet Union, resulted in a decision to modernize the PLA. Two initial steps were taken to promote military Modernisation. First, in 1975, vacant key positions in the military structure and the party CMC were filled. None the less, to ensure party control of the PLA, civilians were appointed to key positions. Deng Xiaoping was appointed Chief of the General Staff, while Zhang Chunqiao was appointed director of the GPD. Second, in 1976, following Premier Zhou Enlai's January 1975 proclamation of the Four Modernisations as national policy, the party CMC convened an enlarged meeting to chart the development of military Modernisation. The military Modernisation program, codified in Central Directive No. 18 of 1975, instructed the military to withdraw from politics and to concentrate on military training and other defence matters. Factional struggles between party moderates and radicals in 1975 and 1976, however, led to the dismissal of Deng from all his posts and the delay of military Modernisation until after the death of Mao Zedong.

The Chinese leadership resumed the military Modernisation program in early 1977. Three crucial events in the late 1970s shaped the course of this program. Firstly the second rehabilitation of Deng Xiaoping, the major civilian proponent of military Modernisation, secondly the re-ordering of priorities in the Four Modernisations, relegating national defence Modernisation from third to fourth place (following agriculture, industry, and science and technology) and thirdly the Sino-Vietnamese border war of 1979. In July 1977, with the backing of moderate military leaders, Deng

Xiaoping reassumed his position as PLA chief of general staff as well as his other party and state posts. At the same time, Deng became a vice chairman of the party Central Military Commission. In February 1980 Deng resigned his PLA position in favour of professional military commander Yang Dezhi, however Deng improved his party CMC position, becoming chairman of it at the Sixth Plenum of the Eleventh Central Committee in June 1981. With enormous prestige in both the military and the civilian sectors, Deng vigorously promoted military Modernisation, the further disengagement of the military from politics, and the shift in national priorities to economic development at the expense of defence.

In 1977-78, military and civilian leaders debated whether the military or the civilian economy should receive priority in allocating resources for the Four Modernisations. The military hoped for additional resources to promote its own Modernisation, while civilian leaders stressed the overall, balanced development of the economy, including civilian industry and science and technology. By arguing that a rapid military buildup would hinder the economy and harm the defence industrial base, civilian leaders convinced the PLA to accept the relegation of national defence to last place in the Four Modernisations. The defence budget accordingly was reduced. Nonetheless, the Chinese military and civilian leadership remained firmly committed to military Modernisation.

The 1979 Sino-Vietnamese border war, although only sixteen days long, revealed specific shortcomings in military capabilities and thus provided an additional impetus to the military Modernisation effort. The border war, the PLA's largest military operation since the Korean War, was essentially a limited, offensive, ground force campaign. China claimed victory, but the war had mixed results militarily and politically. Although the numerically superior Chinese forces penetrated about fifty kilometers into Vietnam, the PLA sustained heavy casualties. PLA performance suffered from poor mobility, backward communications, weak logistics, and outdated weaponry. Inadequate communications, an unclear chain of command, and the lack of military ranks also created confusion and adversely affected PLA combat effectiveness[19].

Contours of PLA Doctrines

Before we get down to study and analyse the war fighting doctrines of the PLA, let's see the definition of Military Doctrine. Wikipedia[20] defines it as:

"Military doctrine is the expression of how military forces contribute to campaigns, major operations, battles, and engagements. It is a guide to action, rather than hard and fast rules. Doctrine provides a common frame of reference across the military. It helps standardize operations, facilitating readiness by establishing common ways of accomplishing military tasks. Doctrine links theory, history, experimentation, and practice. Its objective is to foster initiative and creative thinking. Doctrine provides the military with an authoritative body of statements on how military forces conduct operations and provides a common lexicon for use by military planners and leaders."

The cold war divided the world into two military camps and in spite of the breakdown it has left two models for constructing military doctrines with the students of military strategy. The analysis of the US version of doctrine reveals an emphasis on flexibility of doctrinal application in accordance with different combat environments. Whereas the erstwhile Soviet doctrine tended to emphasize more on the socio – political component than was the case in the West. Another issue was of levels of war and structure of military science. The US officially endorsed the three level concept of war initially in the 1982 edition of army field manual FM-100-5 – Operations and subsequently in 1993 in the field manual of the same name[21]. The doctrine for one level of war is sometimes confused with that applicable to another level. At times the term doctrine is commonly applied to a particular level of war in one military organization frame of reference but it may fall to different framework when applied to other militaries. These issues magnify when we try to fit the concept of doctrine to Chinese military terminologies[22]. Another definition of military doctrine is the limits of elements of national power. It is not a mere wish list but results from compromises reached between various interested parties and from the interaction of many inputs and constrains. It is also the product of the peculiarities of a particular state and a particular army.[23]

The official military strategy of the PRC is embedded in what is known as the 'National Military Strategic Guidelines for the New Period'. This highest level national military guidance has two main components. The first component is operational and it remains the "Active Defence" as adjusted for prosecuting "Local Wars under Conditions of Informationisation. The second component concerns the myriad reforms and Modernisation programs in both "software" and "hardware" the PLA is undergoing. It is generically referred to by the PLA as "army building" or "new period army building".

Most studies on PLA characterize evolution of China's Military Doctrine in a four phase process – Peoples War, Peoples War under Modern Conditions, Local/ Limited war and Local/ Limited war under High technology conditions. Its evolution in a time line is as under

People's War	1927 – 76	Mao Zedong
Peoples War under Modern Conditions	1979 – 84	Deng Xiaoping
Limited/ Local War under Modern Conditions	1985 – 91	Deng Xiaoping
Limited/ Local War under Hi – Tech Conditions	1992 – 2004	Deng Xiaoping
Limited/ Local War under Conditions of Informationisation	2004 onwards	Jiang Zemin

Table 1.1: Time lines Evolution of PLA War fighting Doctrines

People's War

People's war as espoused by Mao during the revolution and anti – Japanese war years is a doctrine for putting people to arms. It is based on recognition of China's inferior military capability to potential adversaries. It counts primarily on the quantitative superiority of China's armed forces, which are predominantly peasantry and improvisation of primitive equipment. To increase the Chinese army's numerical strength, People's War relies heavily on political mobilization of the masses to deliver military might – to provide logistics, intelligence and manpower. Emotion in form of Revolutionary spirit, ideology and nationalism also constitute a driving force. It is comradeship and personal respect rather than bureaucratic procedures and rules that command the army. The Maoist principles – 'retreat when enemy advance; harass when halt; strike when tired; pursue when retreat' and 'lure enemy into the deep' – are essentially products of in-depth defence and strategic interior line operations[24]. In operational terms guerrilla warfare is an integral part of Mao's version of People's War in conduct of total war. For guerrilla strategists, the key to success lies in cumulative effects to chance military balance to their favour before launching counter offensives. As a result the conduct of war is usually made up of lesser actions and an accumulation of more or less random individual victories. In temporal terms, it must be a protracted war.

Transition to Local/ Limited Wars

A quick review of the first few decades after the establishment of the PRC easily lends one to an impression of a country engaging in chronic wars and armed conflicts of varying intensities. At the end of Second World War the iron curtain divided the world into two camps headed by rival super powers: US and the USSR. When North Korea, in June 1950 attacked South Korea for national unification, a new war started, which finally had the UN forces backed by the US and the North Korea and Chinese forces backed by the USSR. Contrary to the doctrines being followed by the adversaries, in the course of three year war, both sides tried to limit the scope of conflict instead of seeking total annihilation of the enemy.

At the beginning of the Korean War, the US observed a policy of "nothing short of total victory in conflict". But the Korean War changed this posture and limited war became the Kennedy administration's official military doctrine. The Russians at first devoted lesser attention to limited wars, which they preferred to, call local wars. By 1960s they began to explore the concept more systematically. Beginning with a study on foreign military research dealing with local wars, the Russian military gradually developed its own work on subject and over time, its publications showed increasing interest and demonstrated greater sophistication on the subject.

Despite their participation in the Korean War and their support to pro Chinese communist factions in South – East Asian insurgencies, the Chinese lagged behind in developing limited war concepts. In fact the Chinese viewed their national survival to be at stake with cold war politics and thus focused on resisting full scale superpower aggression. The Chinese army was thus able to concentrate on fighting a conventional war, in which operational readiness was the key to repelling possible enemy invasion. When the Sino – Soviet alliance broke down and US intensified their intervention in Indo – China in the 1960s, the Chinese military under Mao Zedong decided to purge foreign influence and fall back to a revolutionary Peoples war doctrine. Mass became more important for China to survive a nuclear war and overwhelm enemy troops on ground.

It was only when Deng Xiaoping took over China's leadership in late 1970s that the PLA began to review its military doctrine. In the mid 1980s the PLA began to take keen interest local wars and to incorporate its views on that subject into its own military doctrine. In retrospect, China may have followed the Soviet lead in developing/ following the local war doctrine if not for the breakdown of the Sino – Soviet alliance and the Chinese leaders intensified fear of a nuclear war following that development. Thus for over three decades China had not prepared for local wars, which were at the opposite end to total or peoples war in the spectrum of conflict.

Peoples War under Modern Conditions

China's leaders during the Cultural Revolution perceived the security landscape as capable of fostering wars that could threaten the very survival of the state. The PLA was intended to fight an imminent all out nuclear war – Total war or a local war which would quickly escalate into a total war against a major power and its allies in a nuclear environment. The Chinese answer was the Peoples war. The army was expected to fight protracted wars of attrition rather than relatively short campaigns and conduct defensive rather than offensive operations. All this seemed to justify an infantry oriented military structure. The application of this doctrine sowed the seeds of the 1979 Vietnam disaster. The PLA had been built for a different type of war and was not ready to confront the modern weapons and equipment supplied to Vietnamese by the Russians. Yet even the Sino – Vietnamese conflict was not the initiator for China ultimately effecting the doctrinal change. Rather it should be viewed as a testing ground to demonstrate force readiness and to justify the need to reform when the readiness proved insufficient. The call for new doctrinal innovations within the PLA had in fact preceded that war. People's war failed to answer the challenges posed by new security threats and to achieve national military political objectives. It had resulted in the PLA becoming an obsolescent force. The concept of 'People's War under Modern Conditions' of late 1970s was only a remedy to the defunct Maoist Peoples war. This concept moreover was only reactionary and transitional. While the socio – political meaning of the Peoples war was retained, the military technical element of the new doctrine underwent substantial revision.

PLA having had successfully exploded the nuclear device in 1964, launched IRBM and satellites in 1969[25]. It had demonstrated an indigenous military industrial capacity for producing conventional weaponry. Vietnam was a third world state and Chinese leaders decided to put their countries army to test. China had absolute military power but not equal military readiness. On 17 February 1979 the war started with Chinese employing about 30 infantry divisions attacking along East and West, with strategic targets being Lao Cai, Cao Bang and Lang Son. Twenty eight days later the PLA withdrew from Vietnam, with its official media propagandized stories of valor instead of discussing lessons learnt from the conflict or even disclosing facts about it. It is estimated that approximately 26,000 Chinese soldiers died, 37,000 wounded and 260 taken prisoner. PLA also suffered considerable damage to its equipment and weaponry. A review of the PLA's performance revealed the following[26];

- PLA was operationally and structurally not ready.

- Its soldier's lack of professional education had a cascading effect on availability of qualified and experienced officers.

- The specialized arms were barely proficient.

- A smooth chain of command became a serious issue when war took a heavy toll on junior officers.

- The Maoist legacy, which sacrificed military ranking system for the sake of comradeship, caused the most fundamental problem.

- There were even reports of fratricide within PLA units due to failure to identify friend or foe in the heat of the battle.

- Weakness in PLA's C3 (command, control and communication) war a matter of concern. Their equipment was outdated and susceptible to both jamming and terrain blocking.

- Overall in Vietnam the Chinese were unable to bring the full power of the PLA on to the enemy. PLA Navy was largely irrelevant and the air force operated within the ranges of the PLA SAM cover. Also a major lacuna was identified that the air force had out dated Russian equipment and was largely equipped with interceptors and the air force was unable to initiate offensive action against the Vietnamese.

In addition to the above mentioned deficiencies two other conflicts in 1982 had an over bearing impact on the PLA analysts. These were the Falkland war and the Israeli operations in Lebanon to protect the Galilean settlements.

Deng Xiaoping was the key player in bringing PLA on the track of Modernisation. He initially reassessed China's security environment and revised the threat perceptions that had originally justified Maoist People's war. He assessed that fighting an early and all out war cannot be dismissed, however a major war can surely be postponed. The sudden and localized wars were unpredictable and more threatening. He sanctioned doctrinal debates and openly called for studying People's War under Modern Conditions and finally the doctrine attempted to integrate the old with the new to PLA's maximum advantage. For PLA, Chinese military leaders realized that the future warfare would be combined arms and joint operations. But the indicators pointed to the fact that the Chinese were still preparing to fight protracted wars of attrition to wear down the enemy but would do so by applying revised operational concepts and methods. Hence what came out was revision in the military – technical component of 'People's War under Modern Conditions' which initiated the process of force restructuring, Modernisation, research and development for new and modern weapon systems, combat and live training and command and control[27]. In a summary following were the doctrinal tenants of the military – technical component of 'People's War under Modern Conditions'[28].

- Sudden attacks with ensuing large scale ground invasion perceived as the main threat.

- 'Stand up to enemy' instead of 'luring enemy into the deep'.

- The concentration of forces at strong points to foil enemy attempts at a quick and decisive victory.

- Maintaining a huge army reserve to sustain defence.

- The primacy of positional warfare with supporting mobile and guerilla operations.

- The importance of stabilizing the situation in the initial phase of war in order to provide cover for mobilization.

- Protracted war.

- An emphasis on the offensive at operational and tactical level with a supposedly strategic defensive stance.

Deng Xiaoping was keen to consolidate and boost China's position as a major power in the international security system. 'Peoples War under Modern Conditions' was given substance by PLA's adoption of an active defence strategy which comprised the military technical component of the doctrine. Structurally the PLA had to not only change hardware but also the mix of forces. Operationally it had to upgrade and emphasize training and streamline the chain of command. But an incomplete doctrine led to incomplete reforms. By the mid 80s the Chinese leadership was still preoccupied with fighting a total war. Campaigning was addressed but was never conceived as a separate issue and was always viewed as part of an all out war. Operational offensive and force projection capabilities were still compromised by emphasis on strategic defence.

Local/ Limited War under Modern Conditions

At an enlarged CMC meeting in 1985, the PLA high command bestowed upon the local war doctrine its official blessing and also announced a force reduction of one million servicemen and other structural measures. There after China entered a stage of 'army-building in peacetime', anticipating that any conflict would not be an invasion by an external adversary but localized clashes over contending national interests. While this change in threat perception was beyond doubt significant enough to precipitate doctrinal change, the most important reason for Deg Xiaoping's approach to improve China's military power. In Deng's own words 'even though international situation deteriorates, this force reduction would remain necessary and more necessary because it was a way to enhance military capability'. This change proved critical. The PLA shifted its objective from an initial defensive building posture designed to fight a total war of survival to one more offensive in nature in order to enhance power and assert national interests. During its early years, China could only resort to the strategy of the weak – protracted war. Mass was important to offset initial damage as well as to eventually over whelm the enemy. Local war by contrast emphasizes time and efficiency. It calls for a quick response to seize or regain initiative as conflicts will erupt at short notice and will last for only short time. With this doctrine operational readiness topped the priority list.

Post Vietnam conflict and the study of the Falkland war and Israeli invasion of Lebanon, PLA realized how future wars will be fought by the modern armies and PLA prepared itself to absorb these complexities. From 1945 to 1985 China has fought thirteen local wars and these varied substantially in intensity. Post 1985 enlarged CMC meeting lively discussions emerged in both Chinese internal and external sources. In conclusion it was agreed that Chinese local war consisted of three levels of war: local total war, local limited war and armed conflicts/ sudden incidents. Towards this PLA analysts identified the following characteristics of Chinese local wars[29];

- A threat to China is most likely to be from neighboring countries and regions.

- 'Military struggles' would be integrated with 'political and diplomatic struggles': military operations will be under severe limitations.

- Border conventional warfare would be the major type of war.

- Length and scale of war would vary greatly.

- The war would mainly employ common conventional technologies with little new/ high technologies. Nuclear threat would be minimal but cannot be ruled out.

The doctrines of Local War under Modern Conditions required a very high level of operational readiness, which emphasized the time factor of war and hence was the shift from the structural readiness of the Peoples war era to operational readiness to fight Local wars.

Structural Reforms towards Operational Readiness were as under[30];

- The headquarters of artillery, armour and engineering corps which had been directly under the CMC were placed under GSD.

- The GSD Fourth Department was restored to oversee the strengthening of electronic warfare command and capability.

- New service arms including army aviation corps and electronic warfare units created.

- Eleven military regions reduced to seven.

- The Military Academy, Logistics Academy and Political Academy were merged into National Defence University.

- Army Corps was reorganized into Combined Arms Group Armies some reinforced with electronic warfare units.

- About 4,054 regiment/ division level units and thirty one army corps level units removed.

- One million servicemen were scheduled to be demobilized.

- County level People's Armed Forces Departments (2,592 units) were transferred to the civilian sector.

- As a result of introduction of Non Commissioned Officers system, 76 officer positions filled by rank and file soldiers. The ratio of officer to soldier was reduced from 1:2.45 to 1.33.

- Troops responsible for internal security were transferred to Ministry of Public Security to reinforce People's Armed Police.

- As a result of integration of civil defence with city planning, a quarter of civil air defence projects were transferred to peace time use.

The 1990s witnessed the fruition of a long and arduous process of PLA military doctrinal evolution. The adoption of the Local war doctrine has been a watershed in the process of development on at least two counts. One is that China has been liberated from the minimal security concern of survival to adopt a more assertive posture based on national interests. Chinese military doctrine is today more outward looking. The other relates to the pattern of war as perceived by the Chinese military. To execute this doctrine, the PLA has revamped its structure and operations to harness its offensive capability to conduct military operations outside the Chinese mainland. The restructuring of force mix is producing a mass of higher density. Operational reforms in training and C2 are facilitating movements at higher speeds over greater distances. These are the requirements to be met by any modern army for achieving greater offensive capabilities. However despite the Chinese efforts, doctrinal developments in the 1990s have far outpaced structural and operational reforms.

Local War's under Hi–Tech Conditions and under Conditions of Informationisation

The 1985 doctrinal change signified only the first phase of a longer doctrinal evolution. It was clear departure from the Maoist way of warfare but it did not yet offer a crystallized vision: the intensity, duration and form of localized wars. The Gulf war served to confirm to the Chinese that technology has reshaped the battle field. Not surprisingly, when debates on military technology and RMA were revived in the West in the 1990s, they captured the attention of the PLA. These debates also shocked the PLA that they have not applied RMA to the Chinese situation in a systemic way. The 1991 Gulf War demonstrated the potency of hi – tech weapons and ushered in still another generation of warfare. The war signaled increasing salience of RMA and inevitably triggered yet another round of debate within the PLA. It was not until the ramifications of the Gulf war were assessed and understood that a consensus again emerged with the PLA leadership. It was in 1993 that the period of Local Wars under Hi – Tech Conditions began and which was followed by the period of Local Wars under Conditions of Informationisation. It is not simply 'local wars' that poses challenges to the PLA, but those fought with hi – technology means. Technology becomes niche with which all aspects of military affairs are linked. In line with trends of military reforms since 1980s, the PLA continues to shed numbers and reform its force mix, favouring the so called specialized technical components. It continues to take structural reform measures to enhance operational efficiency. Command automation and training with scientific and technological means and other operational reforms becomes the landmark of the era. At the operational level of war, the military leaders also see a new form emerging. Joint operations, which not long ago replaced combined arms operations as the main force of operation, is about to be taken over by its own refined version – systemic integrated joint operation – to reflect the increasing informatisation of warfare. Accordingly PLA military training and structures began to reform.

The PLA viewed the Gulf war of 1991 with great alarm. This conflict was between the armies prepared to fight two different wars. The Iraqi's planned to fight a typical industrial age or second wave war which relied on mass and aimed at physical destruction. However the US lead coalition waged an information age or third wave war which was centered on knowledge. It was information not so much firepower that made the weapons do their job. The Iraqi military anticipated a Peoples War under Modern Conditions. It was a land power, boasting a quantitative superiority over coalition forces and an advantage in the 'human factor'. Iraqi planning had a lot in common with traditional PLA thinking. The Iraqis counted on an in – depth positional defence and seemingly a protracted war that would deter and if necessary wear down the enemy. They had never renounced the use of chemical weapons and had the advantage of threatening to escalate the conflict by drawing the Israeli's into war. They calculated that they could defeat their superior adversary with inferior equipment but were finally routed. The quick collapse of Iraqi defence was unexpected and galvanized the PLA into fervent studying the latest developments in the West's airpower and air defence that were critical in its success in defeating the Iraqi 'Peoples War' in Gulf.

In early 1993, in a CMC meeting chaired by Jiang Zemin, it was decided that the PLA should be prepared to win 'local wars under hi – technology conditions. The hi – technology local war was defined by PLA researchers as 'armed contests between hi – technology battle systems which control weapon systems of modern production technological level and are capable of conducting war with commensurate operational methods. Such elements as war objectives, targets, war fighting capability, space and time is limited'. After a number of rounds of deliberations and discussions the PLA analysts and researchers listed the following as principles of Local wars under hi – tech conditions.

- Electronic warfare has become an independent campaign.

- Air raid and anti – air raid has become the dominant operational form.

- C3I is the key to military victory.

- Armed helicopters have replaced tanks as main targets in land warfare.

After the Gulf war, the implication of technology on future wars caught the attention of the PLA and since then they have followed the debate on RMA very closely. The impact of RMA on the PLA on its structural and operational readiness is elucidated a little later; however RMA is understood to have systemic effects raising the importance of information, knowledge and intelligence in the whole military system. RMA anticipates more extensive use of informationized and more intelligent platforms, precision guided munitions and command automation system in digitized battle fields and multi dimensional battle space and further development in electronic equipment. Arms and services were expected to be more integrated and new ones to emerge. Military command will be joint and as with the conduct of war – information operations, joint operations and beyond visual range strikes will become basic operations methods. In light of these one understands the call of 'strengthening the army with science and technology' given by Jiang Zemin. In December 2000, the CMC under Jiang Zemin adopted the development strategy of 'accelerating military informatization while undergoing mechanization' – also known as the double historical mission. This acknowledges increasing importance of information in future warfare. While it admits the PLA's technological weakness, it does suggest a strategy of developing pockets of excellence in key aspects to boost over all military capability and also ushers PLA into new age of warfare – under conditions of informatisation. In Chinese conception informationised warfare acknowledges increasing importance of information and related technology in war. It is characterized by enhanced weapon capability and effectiveness by information technology, extensive use of information technology in the battle field especially concerning C4ISR, a redefinition of battle space and form of war participation, especially use of non – professional combat personnel.

Local Wars under the Conditions of Informationalization

The doctrine of fighting Local Wars under the conditions of Informationalization can be explained as NCW with Chinese Characteristics. US DoD defines NCW as "an information superiority-enabled concept of operations that generates increased combat power by networking sensors, decision makers and shooters to achieve shared awareness, increased speed of command, higher tempo of operations, greater lethality, increased survivability and degree of self synchronization. In essence NCW translates information superiority into combat power by effectively linking knowledgeable entities in the battle space".[31]

The US DoD Annual Report to the Congress on Military and Security Developments Involving the People's Republic of China 2010 states that the concept of "informatization" emphasizes the effects of modern information technology on military decision and weapons employment cycles. The term officially entered the PLA's lexicon in 2002 when then Chinese Communist Party (CCP) General Secretary and Central Military Commission (CMC) Chairman Jiang Zemin, in a speech before the 16th Party Congress, referred to the concept as necessary for the PLA's rapid Modernisation and for enabling Integrated Joint Operations. Jiang's address recognized that moving China's military on a path toward informatization would require integrating the entire PLA with common information systems, as well as a new organizational model for war fighting. The PLA formally institutionalized the concept in 2004. PLA analyses of U.S. and coalition operations in Iraq and Afghanistan have re-emphasized the importance of informatization and joint operations.

The PLA is attempting the concurrent pursuit of "mechanization" (application of late 20th-Century industrial technology to military operations) and "informatization" (application of information technology to military operations). As a consequence, and in recognition of the high costs of force-wide refitting with state-of-the-art weapons systems, the PLA is selectively acquiring new generation technologies in some areas, while deferring new acquisitions in others in favour of upgrading older, but capable, systems for networked operations[32].

The Chinese Defence White Paper of 2004[33], gives out the priority and theme of RMA with Chinese Characteristics in very eloquent details. These details are given in the Annexure 1 attached. As per the White Paper, the primary objective of the RMA in the PLA is to make the PLA capable of winning wars under the conditions of informationalization and will be based on China's national conditions and PLA's own state. The priority of the services is Navy, Air force and Second Artillery Force. The time lines were elucidated in the Defence White Paper of 2008[34], where it projected that pursuing the RMA with Chinese characteristics, a solid foundation for the Chinese military will be laid by 2010, accomplish mechanization and make major progress in informationization by 2020 and by mid 21st century reach the goal of Modernisation of National Defence and Armed Forces. The achievements of the PLA undergoing the RMA with Chinese characteristics in informationalization since early 2000s are;

- The focus of the PLA is on informationalization of trans-area systems integration rather than informationization of specific areas, and is on the whole moving well ahead in this stage of comprehensive development.

- Currently, the PLA is aiming at integration, combining breakthroughs in key sectors with comprehensive development, technological innovation with structural reform and development and building of new systems with the modification of existing ones to tap their potentials.

- The PLA continues to enhance systems integration, continues with its efforts to develop and utilize information resources and gradually develop and improve the capability of fighting based on information systems.

- PLA has full-fledged military information systems, with functional command information systems. This integrated military information network came into operation in 2006, resulting in the further improvement of the information infrastructure, basic information support and information security assurance.

- Adequate progress has been made in the building of command and control systems for integrated joint operations, significantly enhancing the capability of battlefield information support.

- IT-based training methods have undergone considerable development and are regularly used as an important training aid.

- Surveying and mapping, navigation, weather forecasting, hydrological observation and space environment support systems have been further optimized for integration and employment.

- A number of information systems for logistical and equipment support have been successfully developed and deployed.

- The PLA educational institutions are a full-scale "digital campuses" with extensive use of information technology and digital connectivity.

- A number of main battle weapon systems have been informationized. The focus is to increase the capability of the main battle weapon systems in the areas of rapid detection, target location, friend-or-foe identification and precision strikes. Some tanks, artillery pieces, ships and aircraft in active service have been informationized, new types of highly informationized combat platforms have been successfully developed, and the proportion and number of precision-guided munitions are on the rise.

- The overall conditions for informationization in the PLA have been greatly enhanced. A leadership, management and consultation system for informationization has been set up, and the centralized and unified leadership for informationization has been strengthened. Theoretical explorations and studies of key practical issues related to informationization have been continuously intensified, medium-and long-term plans and guidance for informationization of the military formulated and promulgated, technical standards revised and refined, and institutional education and personnel training catering to the requirements of informationization strengthened.

The Core Missions of PLA

The New Historic Missions for the PLA were first introduced by President Hu Jintao in 2004[35] and ratified by the Communist Party in 2007[36]. They direct the PLA to carry out four missions:

- Guarantee Chinese Communist Party rule. The PLA is to remain the ultimate backer of the Communist Party.

- Safeguard the strategic opportunity for national development. The PLA is to serve as a powerful defensive force that can deter aggression against China and protect its national sovereignty and territorial integrity so that China may develop economically.

- Safeguard national interests. The PLA must defend China's interests, not only within its land borders, territorial waters, and territorial air space, but also in distant waters, outer space, and in the electromagnetic sphere.

- Play an important role in world peace. China will maintain a defensive military strategy and will participate in United Nations peacekeeping missions and international cooperation on counterterrorism.

Shifts in PLA Doctrines including Organizational and Structural Changes by 2025

War Fighting Doctrines. By the end of this decade or may be earlier a new war fighting doctrine is likely to be introduced in the PLA which will be a shift from the present one of Local Wars Under the Conditions of Informationalization. The primary driver of this shift will be the pace of Modernisation of the PLA, which will achieve the capability of global force projection by the end of 2030s and the growing militaries in the region to include Japan, Australia, Vietnam, Indonesia and India. By 2025s, in the Western Pacific, in addition to US, a resident power in both the Pacific and Indian oceans, Japan, Vietnam and Indonesia will contest the PLA and in the Indian Ocean, India will be a major player, in addition to US and Australia. The other drivers for the shift will be growth of fundamentalism and extremist groups, the non-traditional threats like narcotics, trafficking of humans and animals, climate change, water security etc, impact of social media and internet, management of restive periphery, energy security, economic security and safety and protection of the SLOCs. In addition there will be a greater clarity on the conventions regulating and policing the global commons. The emphasis of the PLA while enunciating the next war fighting doctrine will be on tackling these future threats at a global level and enhancing own capabilities. The timing will depend on the interplay of these drivers and the changing geo strategic and geo political situations in the world. Some indicators that need to be continuously studied and monitored are given under:

- The management of internal situation in China will be an important indicator. An under control and stable internal situation will enable the PLA to look outwards, initiate doctrinal changes and have a doctrine with global ramifications. Vice versa, the PLA will be involved in maintaining peace and harmony internally and the global view will take a back seat and subsequently take more time to implement.

- Organizationally, PLA today is capable of mobilizing over long distances and being effective in an acceptable time frame. However a global army of future cannot base its mobilization of forces including strategic assets on railways and road. A military with global aspirations needs to have strategic air lift capability which can take troops anywhere in the world. For this in addition to heavy and medium airlift aircrafts and allied facilities, the PLA needs a whole system of capabilities and organizations and China needs to develop sustaining and mature relationships with other countries. These developments need to be kept track of and continuously analysed.

- Structurally, PLA has to evolve, especially from command and control point aspect. From the present system the PLA need to shift to an integrated command and control system and accordingly has to implement changes and adjustments in its higher defence organizations. Training proficiency also needs to undergo a change to incorporate the changes required to convert PLA into a military with global outlook.

- The pattern of exercises, incorporation of the services and the command and control structure will indicate towards an impending change in doctrine. Indicators of change will first emerge in the bi annual defence white papers and this will be followed by PLA exercises. Taking example of the Trans Region

Support Operations (TRSO), the concept was first given out in the defence white paper of 2008. PLA started exercising the concept from 2009 onwards in Exercise Stride 2009, Airborne Movement 2009 and Vanguard 2009. The concept was subsequently given out as a theme of further force modernisation in the defence white paper of 2014.

Reorganization of MACs. The reorganization of the MACs will indicate a major shift from not only the way PLA will fight but will also indicate a shift from the established norms especially in the higher ranks of the PLA and the inter se importance of the three services. However there are some impediments in this reorganization are discussed in the subsequent paragraphs:

- PLA has a very special place in the PRC and any leader of the CPC to be a successful President has to have PLA on his right side. In the scenario prevailing today in China, President Xi Jinping has launched the anti graft and mass line campaigns. For these to reach their logical ends, President Xi needs support of the PLA.

- In reorganization of the MACs, for the PLA at stake are fourteen four star general ranks. If there is a reorganization, not only the ranks at the highest level, but its fall out on all the ranks of the PLA officer cadre and the existing promotion ladder will have to be managed by the Chinese leadership.

- The PLAA has always been the dominant component of the PLA. The CMC till the present one always was dominated by the PLAA with one member each of the PLAAF and PLAN. Its only in the present CMC, PLAAF has two representatives and PLAN has one. All MAC commanders till date are from the PLAA with PLAAF and PLAN commanders, though not subservient but not equal for sure. Reorganization of the MACs will alter this status quo.

- The logistics, recruitment/ conscription and the mobilization of the PLA is organized by and the responsibility of the Military Districts (MD). The MDs are under the command of the MACs and geographically also form part of the MAC. This has functioned well for the PLA for many years and interests have developed along the chain of command. The reorganization of MACs will not only affect these interests but will have an impact on the Chinese society.

- In near future, if both Nanjing and Guangzhou MACs or even one of them gets commander from the PLAN, this for sure be an indicator of impending reorganization of the MACs

Downsizing of the PLA. Size of PLA as brought out earlier, will depend on the internal situation of PRC and the changing geo strategic and geo politics of the region and the world. Downsizing of the PLAA will continue and one may see this trend once or twice. Even today there are unconfirmed inputs that point towards downsizing of PLA from 2.33 million to 2.03 million and a confirmation of this may be there in the next Defence White Paper of the PLA due by end 2014.

Endnotes

1   Opening address at the First Plenary Session of the Chinese People's Political Consultative Conference (September 21, 1949)

2   History extracted from the PLA History as given in the website of GlobalSecurity.org   http://www.globalsecurity.org/military/world/china/pla-history.htm accessed on May 11, 2013.

3   Ibid.

4   Mark A Rayn, David M Finkelstien and Michael A McDevitt, Chinese Warfighting: The PLA Experiences Since 1949, M E Sharpe, New York, pp. 24-25.

5   Ibid, p. 26.

6   "Chinese People's Liberation Army", CPC Encyclopedia, www.cpcchina.org/2011-10/19/content_13994372.htm .

7   Wei Wang,"The Chinese People's Liberation Army", China Intercontinental Press, Beijing, 2012, p. 97.

8   Ibid.

9   Mark A Rayn, David M Finkelstien and Michael A McDevitt, Chinese Warfighting: The PLA Experiences Since 1949, M E Sharpe, New York, pp. 137-139.

10   Shiping Zheng, Party vs State in Post-1949 China: The Institutional Dilemma, Cambridge University Press, Cambridge, UK, 1997, p. 108.

11 Sino-Soviet Relations, www.countrystudies.us/china/128.htm, accessed on 13 May 2014.

12 Lt Gen J S Bajwa, "Modernisation of the Chinese PLA: From Massed Militia to Force Projection", Lancer Publications, New Delhi, 2013, p.42.

13 Ralph L Powell, "Commissars in the Economy:"Learn from the PLA" Movement in China", Asian Survey, 5(3) March 1965, pp. 126-127.

14 Lt Gen J S Bajwa, "Modernisation of the Chinese PLA: From Massed Militia to Force Projection", Lancer Publications, New Delhi, 2013, p.50.

15 Ibid, p. 52.

16 Xuezhi Guo, China's Security State: Philosophy, Evolution and Policies, Cambridge University Press, Cambridge, UK, 1997, pp. 126-129.

17 Ellis Joffe, "Party Army Relations in China: Retrospect and Prospect", The China Quarterly No 46, June 1996, p. 310.

18 Mark A Rayn, David M Finkelstien and Michael A McDevitt, Chinese Warfighting: The PLA Experiences Since 1949, M E Sharpe, New York, p. 225.

19 Ibid, pp. 235-236.

20 http://en.wikipedia.org/wiki/Military_doctrine

21 FM 100-5 – Operations, 1993, p. 1-3, accessed from http://www.fprado.com/armoursite/US-Field-Manuals/FM-100-5-Operations.pdf on May 13, 2014.

22 Ka Po Ng, "Interpreting China's Military Power, Doctrine makes readiness", Routledge, New York, US, 2005, pp17 – 18.

23 Ibid, pp 22 – 23.

24 Mark A Rayn, David M Finkelstien and Michael A McDevitt, Chinese War fighting: The PLA Experiences Since 1949, M E Sharpe, New York, US, p. 26.

25 DF-3A/ CSS-2, fas.org, www.fas.org/nuke/guide/china/theatre/df-3a.htm, accessed on May 13, 2014.

26 Ka Po Ng, "Interpreting China's Military Power, Doctrine makes readiness", Routledge, New York, US, 2005,pp. 65-70.

27 Ibid, pp 71 – 74.

28 Ibid, p. 71.

29 Ibid, p. 91.

30 Ibid, p. 96.

31   Alberts David S, Garstka John J. and Stein Frederick P., *Network Centric Warfare: Developing and Leveraging Information Superiority,* DoD C4ISR Cooperative Research Program, Washington DC, 2000, p. 2.

32   Annual Report to the Congress, *Military and Security Developments Involving the People's Republic of China 2010*, US DoD, 2010, p. 3.

33   Information Office of the State Council, *"China's National Defence in 2004"*, November 2004, Beijing, p.7 to 16.

34   Information Office of the State Council, *"China's National Defence in 2008"*, January 2009, Beijing, p.7.

35   2005: Chairman Hu Jintao defines historical mission of PLA in new period at http://english.chinamil.com.cn/site2/special-reports/2008-11/27/content_1609747.htm (Accessed June 10, 2014)

36   Pollpeter Kevin, "Controlling the Information Domain: Space, Cyber and Electronic Warfare", in Ashley J. Tellis & Travis Tanner (ed.), Strategic Asia 2012-13: China's Military Challenge, National Bureau of Asian Research, Washington DC, 2012.

2

PLA Army (PLAA) Modernisation

"China does not now face a direct threat from another nation. Yet, it continues to invest heavily in its military, particularly in programs designed to improve power projection. The pace and scope of China's military build-up are, already, such as to put regional military balances at risk....In the future, as China's military power grows, China's leaders may be tempted to resort to force or coercion more quickly to press diplomatic advantage, advance security interests, or resolve disputes."

(Annual Report To Congress, The Military Power of the People's Republic of China 2005, July 19, 2005)

Introduction

PLAA today also forms the numerical bulk of PLA forces, 1.6 million out of a total active duty force of 2.25 million. It is often said that of these, somewhere between 35–50 per cent of the ground forces are at full strength and readiness, and are ready to deploy for combat (this likewise means that half or more are not ready for deployment)[1]. Dennis J Blasko[2] gives a three tiered pyramid explaining the relationship of the three schools of military thoughts prevalent in PLA today: Peoples War, Local War and the RMA School. These three schools are reflected in the PLA's doctrinal development, equipment and scenario planning. The base of the pyramid consists of the Peoples War School, where military thought of Mao Zedong provides theoretical foundation. At present 70 per cent of the PLA is best suited to fight a People's War and is equipped with weapons and equipment designed in 1950's and 60's. The tactics these units practice are similar to those used in the War against Japan, the War of Liberation, the Korean War, and the 1979 conflict with Vietnam. The second tier of the PLA pyramid is the Local War School which forms maybe 25 per cent of all PLA. Deng Xiaoping provided the critical strategic direction for this school. Local War is understood to be a limited war on the periphery of China and will be short but intense, utilizing advanced technology weapons, with units fighting in joint and combined arms efforts. It envisions an element of force projection (i.e., the ability to transport combat forces beyond China's borders), but by definition is regional, not global, in nature. China usually regards Local War as its "next war"; the Persian Gulf War is often a point of reference for this school. However PLAA has no combat experience in this type of conflict. It is this portion of PLA which is expected to grow in future as the Peoples War segment shrinks. The RMA School is at the top of the pyramid and is represented by a very small portion of the PLA. It is these elements which are regularly referred to as the "pockets of excellence" in large no of professional literature. No Chinese leader has given a direction to this concept and hence lack of focus and high level vision of future war may slow down the development of many of the concepts currently being explored by thinkers at relative lower ranks.

Article 22 of China's Law on National Defence adopted on March 14, 1997 states *"The Armed Forces of the People's Republic of China are composed of the active and reserve units of the Chinese People's Liberation Army (PLA), the Chinese People's Armed Police Force (PAP), and the People's Militia."*

The Defence White Paper released in 2013 generally gives the composition and lays down the direction in which the PLAA will transform and modernize in the times ahead. It says

"The PLA Army (PLAA) is composed of mobile operational units, border and coastal defence units, guard and garrison units, and is primarily responsible for military operations on land. In line with the strategic requirements of mobile operations and multi-dimensional offense and defence, the PLAA has been reoriented from theater defence to trans-theater mobility. It is accelerating the development of army aviation troops, light mechanized units and special operations forces, and enhancing building of digitalized units, gradually making its units small, modular and multi-functional in organization so as to enhance their capabilities for air-ground integrated operations, long-distance maneuvers, rapid assaults and special operations. The PLAA mobile operational units include 18 combined corps, plus additional independent combined operational divisions (brigades), and have a total strength of 850,000. The combined corps, composed of divisions and brigades, are respectively under the seven military area commands (MACs): Shenyang (16th, 39th and 40th Combined Corps), Beijing (27th, 38th and 65th Combined Corps), Lanzhou (21st and 47th Combined Corps), Jinan (20th, 26th and 54th Combined Corps), Nanjing (1st, 12th and 31st Combined Corps), Guangzhou (41st and

42nd Combined Corps) and Chengdu (13th and 14th Combined Corps)."[3]

An analysis of the above clearly indicates the following trends of PLAA Modernisation

- PLAA is going to mould itself into a **modern military force** which will be capable of **joint and mobile operations** in a digitized environment with trans – theatre mobility leading to a full spectrum expeditionary capability.

- The focus of modernisation will be on **aviation, mechanization and Special Forces**.

- The units will be **small, modular and multifunctional**, there by overcoming the present dispensation of role i.e. RRF or Light Mechanized etc and terrain i.e. jungle or high altitude or amphibious. The new form of units will be able to operate in almost all types of terrains and will be able to react to various contingencies with the similar efficiency. Clear indicators regarding the pace of this transformation will emerge in future as PLAA trains, educates, recruits and equips its army.

- The move to call the MRs as MACs, GAs as CCs and Brigades as Combined Operational Divisions is a clear indicator to the trend where the PLAA wants to be perceived as a modern and integrated army capable of operating in a digitized and joint environment, a theme which will be central to future PLAA modernizing and force structuring.

- As far as Indian armed forces are concerned, by calling the PLA brigades as Combined Operational Division, the percept that the PLA brigades are similar to the Indian brigades has been cleared. The size of a PLA brigade is, if put very generally, is about half the size of Indian army division.

Trends of PLAA Force Modernisation

A comprehensive view of PLAA's Modernisation Trends can be obtained from a study of authors i.e. Richard Fisher's book 'China's Military Modernisation-Building for Regional and Global Reach,' David Shambaugh 'Modernizing China's Military' and Dennis J Blasko's 'The Chinese Army Today – Tradition and Transformation for the 21st Centuary.'

Besides piecemeal inputs can be accessed from the Pentagon's reports to US Congress and other open source material. The key points of military Modernisation are summarised in the table below:-

Table 2.1 Key Points in PLAA Moderisation	
Capabilities	**Trends**
C4I2SR	• Increased deployment and employment of UAVs and UCAVs. • Development of Chinese information operations able to degrade sophisticated enemy intelligence, surveillance, and reconnaissance systems.
Integrated Joint Opera-tions	• Development of better air defence capabilities, including integration of more advanced surface-to-air missiles like the S-300 PMU-2, HQ-9, S-400 and the HQ-19. • The ability to coordinate and execute multi-service and joint operations in the various battle space dimensions. • Improvements in communications architectures that enable war zone commanders to coordinate the movements and actions of major units across military region boundaries. • An increase in the number of command post exercises in which officers from different military regions and services practice joint command-and-control activities.
Precision Strikes	• Improvement in targeting technologies, especially over the horizon targeting. • Development and enhanced use of precision-guided munitions (PGMs). • Development of stealthy, long-range cruise missiles.
Combat Support	• Improvements in the joint logistics system. • Development of in-flight refueling and airborne command and control capabilities • Moderate increase in airlift ability. • Moderate increase in sealift capabilities.
Training	• Improved execution of training exercises that involve joint ground and air units.

Certain Other Indicators	<ul><li>Reorganization of the MACs.</li><li>A concerted effort to enhance the froce projection capabilities of the PLAA in conjunction with the PLAAF and PLAN. The capapbilities will include the amphibious, airborne, airlift and helilift capabilities both with in China, in its periphery and over long distances away from its border.</li></ul>

In this part we explore the state of China's military Modernisation program till 2025, assess how much progress the PLA Army (PLAA) has made, elucidate its principal trends and trajectories and identify important indicators to monitor in the future.

Most of the studies of PLA Modernisation in the past have been confined to the factors and inputs that are determining China's military Modernisation program. These factors and inputs have been the focus in part because the parameters of PLA capabilities are fairly well known and can be summarized rather succinctly and in part because it is more analytically important to identify the variables that are shaping and driving the program. Knowing the weapons systems inventory of a military is only of limited value—more important is to understand the strategy, doctrine, politics, perceptions, contingencies, technologies, manpower, training, logistics, and other "software" factors that shape the use of force. In the final calculation, how a military fights and wages war depends more on the latter factors than on the former. Half of ground force deployments (approximately 800,000) of the PLA remain concentrated in North, North East, and Eastern China. In an attempt to reduce costs, improve readiness, and eliminate redundancies, the ground forces have undergone a substantial streamlining and downsizing in recent years. Combined Corps (comprised of between 30,000 – 65,000 troops) remain the center of the main force deployments, but in an attempt to downsize to the division and brigade levels (with approximately 12,000–15,000 personnel), they are undergoing structural changes. This streamlining is intended to improve mobility, jointness, and combined-arms capabilities. A substantial increase in transport helicopters has also contributed to the rapid reaction capabilities of the ground forces. The education levels of ground force officers, NCOs and conscripts have also increased. While the ground forces have received lesser priority relative to other services in recent years, they remain the backbone of the PLA. The PLA ground forces field a full range of equipment, including tanks, armoured personnel carriers, artillery, surface-to-surface and surface-to-air missiles, and helicopters and unmanned aerial vehicles. While the numbers of inventory are large, the quality remains very uneven and much of the hardware remains antiquated. For example, of the 7,050 main battle tanks, 5,100 are of 1950s–1980s vintage. The remaining 1,950 are composed of the T-96, T-96A, T-98A and T-99 main battle tanks. The T-99 series tanks have fairly advanced fire control systems, laser range-finders, all-weather computerized control, hardened composite armour, and an engine with improved speed and power-to-weight ratio. The ground forces also possess a huge number of towed artillery (14,000), antitank weapons (7,200), air defence guns (7,700), and other conventional land systems.

The analysis of the force structure and the military balance[4] further strengthens these aspects. If we take the example of the PLA Armoured formations this aspect is further highlighted.

Table 2.2: Analysis PLAA Armoured Formations

Equipment	Total Numbers	Number per mechanized brigade/ armoured brigade	Number of Brigades	Remarks
Main Battle Tank: Type 96/ 96A	1,500	Mechanized Brigade: 99 Tanks	Approx 19 Mechanized Brigades	PLA Army has: • Armoured Divisions – 9 • Armoured Brigades – 9
Type 98A/ 99	450	Armoured Brigade: 297 Tanks	Or Six Armoured Brigades	• Mechanized Divisions – 9 (including two Amphibious Divisions)
Total	1,950			• Mechanized Brigades – 8 (including two Mechanized Regiments)

This clearly that there are only adequate numbers of modern fourth generations tanks to equip their 'elite' formations or the brigades. An analysis of other arms like the mechanized infantry, artillery, aid defence and other supporting arms also indicate to the same trend.

Table 2.3: Analysis Transformation of Group Armies

Ser No	Formations	1985	1990	1995	2014 (As per Military Balance 2014)[5]
(a)	MRs/ MACs	11	7	7	7
(b)	GAs/ CCs	35 Field Armies	24 GA	24 GAs	18 CCs
(c)	Infantry/ Motor-ized/ Mechanized Divisions	118	80	73	25
(d)	Infantry/ Motor-ized/ Mechanized Brigades	0	0	0	29
(e)	Armoured Divi-sions	13	10	11	9
(f)	Armoured Bri-gades	0	0	0	9
(g)	Airborne Divisions (Under PLAAF)	3	3	3	3

Force Mobility

PLA ground forces were focused till 2010 on internal issues such as border protection, disaster relief and training for internal-security operations. Improving the ability of the ground forces to project power domestically, by using military and civilian transport infrastructure, is crucial to force development. During exercises such as October's Duty Action 2010, light mechanized units have moved ever-longer distances by rail and road, while being subjected to simulated attack and electronic interference. This is seen by the PLA as the best way of testing the resilience of its ground units in fighting a 'modem war under informationized conditions', a primary goal mentioned in all the Defence White Papers published after 2004. Meanwhile, the use of civil-military transport also allows the PLA to deploy some of its key formations faster, thus enhancing its ability to protect borders and to consolidate control in remote regions. These developments are noteworthy given the difficulties in using fixed-wing airlift in the Himalayan border region, and the lack of growth in airlift capabilities. A 2005 agreement to purchase 34 Il-76 transport aircraft and four Il-78 Midas AAR from Russia remained frustrated by negotiations over contract value over a long time[6]. Civilian transport networks have been integrated into the PLA's logistical infrastructure and can provide troop transport with little notice. This has dramatically increased the mobility of PLA ground formations inside China, and is particularly important in southern Tibet, where the Qinghai-Tibet railway can rapidly move units from the Chengdu military region to Lhasa or to the disputed Sino-Indian border. These units would be used to reinforce the relatively few formations based in Tibet. PLA units in the southern military regions have also augmented their amphibious capabilities with large civilian ships, such as roll-on roll-off ferries, which would be helpful in any amphibious scenario.

Smaller, More Mobile Forces

To improve rapid-reaction capabilities, most light mechanized units have been re-equipped with wheeled infantry fighting vehicles. Supporting combat units have also been issued new weapons, including the self-propelled PTL-02 100 mm assault gun and Type 95 SPAAG, which are now in service with antitank and air-defence battalions respectively. Although integrating new equipment has not always proved straightforward, the PLAA has increasingly worked closely with industry when bringing new systems into service. The move towards smaller, more mobile forces continues a long-running doctrinal and organizational shift in the army. In the late 1990s, the PLA began disbanding dozens of heavy, divisions and creating smaller brigades - producing a core of more mobile mechanized and motorized formations. This process was halted in 2003 because of a lack of officers experienced in commanding such formations. However, the PLAA has since experimented with 'special-mission battalions', usually an infantry battalion of a mechanized or motorized brigade specifically trained for ad-hoc quick-reaction missions and rapid deployment. This trend towards special-mission battalions has intensified since the publication of the new military training and evaluation manual in 2006, with the formation of 'combined battalions', i.e., battalion battle groups formed from company-sized units from up to a dozen different branches of the armed forces.

According to the PLAA, this reorganization within the army is significantly improving operational flexibility, although China's continuing opacity in the military sphere complicates any objective assessment of the PLA's success in adapting to the new doctrine. The importance of the battalion-battle-group concept is highlighted by the use of elite units in the new combined battalions. Because they can adapt more easily than other, less-trained units of the PLAA, units such as the elite 58th Mechanized Brigade (Jinan Military Region) and the 1st Armoured Division (Nanjing Military Region) are among those actively training in the new battalion battle groups[7].

The PLAA's new operational concept was tested during Peace Mission 2010 in Kazakhstan, in which a battalion battle group formed from light mechanized units of the Beijing Military Region took part. Armoured, mechanized and aviation units were supported by a range of artillery, assault-gun, electronic- warfare and logistical personnel. The PLA appeared more satisfied with this arrangement than with previously tested brigade-level formations, and has instigated widespread combined-battalion training. It is believed that the idea is not to make these combined battalions permanent, but merely to allow units within the PLAA to form ad hoc battle groups as needs dictate, allowing for a more 'modular' army. This trend continues and in the Peace Mission 2014 a similar concept was exercised. The newly trained battalion commanders who emerge from the process should also mean that the process of downsizing most of the remaining 38 divisions to brigades can in future be resumed.

The requirement for smaller operational formations stems from the changing security environment in which the PLAA finds itself. Just as the brigade experiments were a response to the end of the Soviet threat the battalion battle group is largely a response to concern over domestic and other low-intensity threats. Smaller and faster-reacting units can better respond to domestic instability, as well as fight counter-insurgency campaigns. The move towards lighter and more mobile forces may slowly degrade the PLA's doctrinal preference for heavy armour. Currently, all PLAA divisions retain an armoured regiment, even light rapid-reaction units such as the 149th Light Mechanized Division in the 13th Group Army in Chengdu Military Region. Nevertheless, regional differences in the employment of armour suggest that military chiefs recognize that tanks are not necessarily suited to every contingency, with light troops being used in areas with developed road networks and tracked units in difficult terrain. For example, combined battalions in the Jinan Military Region usually include light mechanized units from the 58th Brigade with some tanks in support, while those in the Nanjing Military Region - opposite Taiwan - include many tanks and tracked infantry fighting vehicles. These have been analysed in detail in the later part of this chapter.

PLAA Brigade Organizations

After the PLAA started converting/ disbanding divisions in favour of brigades in late 1990s, the mechanized brigade was seen as the main operational unit for the future (until the view gained traction that these doctrinal changes were best developed through creating flexible and modular battalions). There are now seven mechanized brigades in PLA out of which five – 23, 188 (27 GA), 58 (20 GA), 190(39 GA), and 139 (39 GA) – are considered elite formations[8]. PLA mechanized brigades are divided into light and heavy: former equipped with wheeled APCs such as Type 92/92A; the latter with tracked IFVs/ APCs such as Type 63/89. All mechanized brigades have at least one battalion of tanks, while certain elite formations have two battalions each. These tank battalions are accompanied by three battalions of infantry. All PLAA battalions follow 3+3 organization of three companies with three platoons. There is also a scaled down Artillery Regiment under the direct control of the Brigade Commander. This consists of one battalion each of 122mm and 152mm howitzers and one company of 122mm MRLs[9]. The approximate size of PLAA units and formations age as under[10]:

Table 2.4: Approximate Strength of PLAA Formations and Units

Unit Type	Division	Brigade	Regiment	Battalion
Infantry	10,000-12,000	5,000-6,000	2,800	700
Armour	10,000	2,000	1,200	175
Artillery	5,000-6,000	2,200	1,100	275
AAA	5,000	2,000	1,100	250
Airborne (PLAAF)	8,000-9,000	5,000-6,000	2,300	700

Brigadization of the PLAA: Trends[11]

The trends of transformation of the CCs of the various MACs are as given under:-

Table 2.5: Trends of Transformation MAC

Beijing MAC - Trend of Transformation Three Combined Corps (CC) (27, 38 & 65)				
27 CC (2009)	**27 CC (2011)**	**27 CC (2013)**	**27 CC (2014)**	**Remarks**
• 1 Armd Bde • 2 Mech Bde • 2 Mot Bde • 1 Arty Bde • 1 AD Bde • 1 Engr Regt	• 1 Armd Bde (OPFOR) • 2 Mech Inf Bde • 2 Mot Inf Bde • 1 Arty Bde • 1 AD Bde • 1 Engr Regt	• 1 (OPFOR)* Armd Bde • 2 Mech Inf Bde • 2 Mot Inf Bde • 1 Arty Bde • 1 AD Bde • 1 Engr Regt	• 1 Armd Bde • 2 Mech Inf Bde • 2 Mot Inf Bde • 1 Arty Bde • 1 AD Bde • 1 Engr Regt	• Defensive Combined Corps. • All formations were brigadized in 2009 • One armoured brigade earmarked for Out of Area contingencies removed from the orbat
38 CC	**38 CC**	**38 CC**	**38 CC**	
• 1 Armd Div • 2 Mech Div • 1 Arty Bde • 1 AD Bde • 1 Engr Bde • 1 Avn Regt	• 1 Armd Div • 2 Mech Div • 1 Arty Bde • 1 AD Bde • 1 Engr Bde • 1 Avn Regt	• 1 Armd Div • 2 Mech Div • 1 Arty Bde • 1 AD Bde • 1 Engr Bde • 1 Avn Bde	• 1 SF Unit • 1 Armd Div • 2 Mech Div • 1 Arty Bde • 1 AD Bde • 1 Engr Regt • 1 Avn Bde	• Offensive Combined Corps. • Armoured & mechanized divisions yet to be reorganized. • A SF Unit and EW Regt now on the orbat of the CC.

65 CC (2009)	65 CC (2011)	65 CC (2013)	65 CC (2014)	
• 1 Armd Div • 1 Mot Div • 1 Arty Div • 2 Mot Bde • 1 AD Bde • 1 Engr Bde	• 1 Armd Div • 1 Mech Inf Div • 1 Mot Inf Bde • 1 Arty Bde • 1 AD Bde • 1 Engr Regt • 1 Avn Regt	• 1 Armd Div • 1 Mech Inf Div • 1 Mot Inf Bde • 1 Arty Bde • 1 AD Bde • 1 Engr Regt • 1 Avn Bde	• 1 Armd Bde • 1 Mech Inf Div • 1 (OPFOR)* Mech Bde • 2 Mot Inf Bde • 1 Arty Bde • 1 AD Bde • 1 Engr Regt	• Defensive Combined Corps. • Armoured & mechanized divisions yet to be reorganized • *One mechanized infantry brigade earmarked for Out of Area contingencies
Beijing: 2 Mech Regt	Beijing: 2 Survey Div			
MR or MD Units	**MR or MD Units**	**Other Forces**	**Other Forces**	
• 1 Engr Bde (PKO) Regt • 1 Spec Ops Unit • 1 AAA Bde	• 1 Mot Inf Bde • 1 Spec Ops Unit • 1 ADA Bde	• 2 (Beijing) Gd Div • 1 Mot Inf Bde • 1 SF Unit • 1 ADA Bde	• 2 (Beijing) Gd Div • Avn Regt	

Shenyang MAC - Trend of transformation Three Combined Corps (CC) (16, 39,49)				
16 CC (2009)	**16 CC (2011)**	**16 CC (2013)**	**16 CC (2014)**	
• 1 Armd Div • 2 Mot Div • 1 Mot Bde • 1 Arty Bde • 1 AD Bde • 1 Engr Bde	• 1 Armd Div • 2 Mot Inf Div • 2 Mot Inf Bde • 1 Arty Bde • 1 AD Bde • 1 Engr Bde	• 1 Armd Bde • 1 Mech Inf Bde • 2 Mot Inf Div • 2 Mot Inf Bde • 1 Arty Bde • 1 AD Bde • 1 Engr Regt	• 1 Armd Bde • 3 Mech Inf Bde • 2 Mot Inf Div • 1 Arty Bde • 1 AD Bde • 1 Engr Regt	• Defensive CC • Brigadization, except two motorized divisions, has been completed.
39 CC	**39 CC**	**39 CC**	**39 CC**	
• 1 Armd Div • 1 Mech Div • 1 Mot Div • 1 Mech Bde • 1 Arty Bde • 1 AD Bde • 2 Avn Regts	• 1 Armd Div • 1 Mech Inf Div • 1 Mot Inf Div • 1 Mech Bde • 1 Arty Bde • 1 AD Bde • 1 Avn Regt	• 1 Armd Bde • 1 Mech Inf Div • 1 Mot Inf Div • 1 Mech Bde • 1 Arty Bde • 1 AD Bde • 1 Avn Regt	• 1 SF Unit • 1 Armd Bde • 1 Mech Inf Div • 2 Mech Inf Bde • 1 Mot Inf Div • 1 Arty Bde • 1 AD Bde • 1 Avn Regt • 1 Engr Regt • 1 EW Regt	• Offensive CC • The mechanized and motorized infantry divisions left to be reorganized.
40 CC	**40 CC**	**40 CC**	**40 CC**	
• 1 Armd Bde • 3 Mot Bde • 1 Arty Bde • 1 AD Bde • 1 Engr Regt	• 1 Armd Bde • 2 Mot Inf Bde • 1 Arty Bde • 1 AD Bde • 1 Engr Regt	• 1 Armd Bde • 2 Mot Inf Bde • 1 Arty Bde • 1 AD Bde • 1 Engr Regt	• 1 Armd Bde • 3 Mot Inf Bde • 1 Arty Bde • 1 AD Bde • 1 Engr Regt	• Defensive CC • All formations were brigadized in 2009

MR/ MD Units	MR/ MD Units	MR/ MD Units	MR/ MD Units	
• 1 EW Regt • Spec Ops Unit	• 1 Mot Inf Bde • 1 EW Regt • Spec Ops Unit	• 1 Mot Inf Bde • 1 EW Regt • SF Unit	• -nil-	
Jinan MAC - Trend of transformation **Three Combined Corps(CC) (20, 26 & 54)**				
20 CC (2009)	**20 CC (2011)**	**20 CC (2013)**	**20 CC (2014)**	
• 1 Armd Bde • 1 Mech Bde • 1 Mot Bde • 1 Arty Bde • 1 AD Bde • 1 Engr Regt	• 1 Armd Bde • 1 Mech Inf Bde • 1 Mot Inf Bde • 1 Arty Bde • 1 AD Bde • 1 Engr Regt	• 1 Armd Bde • 1 Mech Inf Bde • 1 Mot Inf Bde • 1 Arty Bde • 1 AD Bde • 1 Engr Regt	• 1 Armd Bde • 2 Mech Inf Bde • 1 Arty Bde • 1 AD Bde • 1 Engr Regt	• Offensive CC • All formations were brigadized in 2009
26 CC	**26 CC**	**26 CC**	**26 CC**	
• 1 Armd Div • 3 Mot Bde • 1 Arty Bde • 1 AD Bde • 2 Avn Regt	• 1 Armd Div • 3 Mot Inf Bde • 1 Arty Bde • 1 AD Bde • 1 Avn Regt	• 1 Armd Div • 3 Mot Inf Bde • 1 Arty Bde • 1 AD Bde • 1 Avn Regt	• 1 SF Unit • 1 Armd Bde • 1 Mech Inf Bde • 3 Mot Inf Bde • 1 Avn Regt • 1 Arty Bde • 1 AD Bde • 1 Engr Regt • 1 EW Regt	• Offensive CC • All formations were brigadized in 2009 less the armoured division.

54 CC (2009)	54 CC (2011)	54 CC (2013)	54 CC (2014)	
• 1 Armd Div • 1 Mech Div (RRU) • 1 Mot Div (RRU) • 1 Arty Bde • 1 AD Bde • 1 Avn Regt	• 1 Armd Div • 2 Mech Div (RRU) • 1 Arty Bde • 1 AD Bde • 1 Avn Regt	• 1 Armd Bde • 2 Mech Div (RRU) • 1 Mech Inf Bde • 1 Arty Bde • 1 AD Bde • 1 Avn Regt	• 1 Armd Bde • 2 Mech Div (RRU) • 1 Mech Inf Bde • 1 Arty Bde • 1 AD Bde • 1 Avn Regt • 1 Engr Regt	• Offensive CC • All formations were brigadized in 2009 less the RRU.
MR or MD Units	**MR or MD Units**	**Other Forces**	**Other Forces**	
• 1 EW Regt • 1 Spec Ops Unit • 2 Pontoon Br Regt	• 1 EW Regt • 1 Spec Ops Unit	• 1 EW Regt • 1 SF Unit		

Nanjing MAC - Trends of Transformation
Three Combined Corps (CC) (1, 12 & 31)

1 CC	1 CC	1 CC	1 CC	
• 1 Armd Div • 1 Amph Div • 1 Mot Bde • 1 Arty Bde • 1 AD Bde • 1 Engr Regt	• 1 Amph Div • 1 Arty Div • 1 Armd Bde • 1 Mot Inf Bde • 1 AD Bde • 1 Engr Regt • 1 Avn Regt	• 1 Amph Mech Inf Div • 1 Arty Div • 1 Armd Bde • 1 Mot Inf Bde • 1 AD Bde • 1 Engr Regt • 1 Avn Regt	• 1 Armd Bde • 1 Amph Mech Inf Div • 1 Mech Inf Bde • 1 Mot Inf Bde • 1 Avn Bde • 1 Arty Div • 1 AD Bde • 1 Engr Regt • 1 EW Regt	• Amphibious CC • All formations are brigadized which are less those likely to be employed in the Taiwan contingency.

12 CC (2009)	12 CC (2011)	12 CC (2013)	12 CC (2014)	
• 1 Armd Div • 3 Mot Bde • 1 Arty Bde • 1 AD Bde • 1 Engr Regt	• 1 Armd Div • 3 Mot Inf Bde (1 RRU) • 1 Arty Bde • 1 AD Bde • 1 Engr Regt	• 1 Armd Div • 3 Mot Inf Bde (1 RRU) • 1 Arty Bde • 1 AD Bde • 1 Engr Regt	• 1 Armd Div • 2 Mech Inf Bde • 2 Mot Inf Bde (1 RRU) • 1 Arty Bde • 1 AD Bde • 1 Engr Regt	• Amphibious CC • All formations less the armoured division were brigadized in 2009
31 CC	**31 CC**	**31 CC**	**31 CC**	
• 2 Mot Div (1 (86th) RRU) • 1 Amph Armd Bde • 1 Mot Bde • 1 Arty Bde • 1 AD Bde • 1 Avn Regt	• 2 Mot Inf Div (1 RRU) • 1 (Amph) Armd Bde • 1 Mot Inf Bde • 1 Arty Bde • 1 AD Bde • 1 Avn Regt	• 2 Mot Inf Div (1 RRU) • 1 (Amph) Armd Bde • 1 Mot Inf Bde • 1 Arty Bde • 1 AD Bde • 1 Avn Regt	• 1 SF Unit • 1 (Amph) Armd Bde • 2 Mot Inf Div (1 RRU) • 1 Mot Inf Bde • 1 Avn Regt • 1 Arty Bde • 1 AD Bde • 1 Engr Regt • 1 EW Regt	• Amphibious CC • All formations were brigadized in 2009. The motorized infantry division left to be reorganized is a RRU.
MR or MD Units	**MR or MD Units**	**Other Forces**	**Other Forces**	
• 1 Avn Regt • 1 Spec Ops Unit	• 1 SSM Bde • 1 Spec Ops Unit	• 1 SF Unit		

Trends of Transformation Guangzhou MAC Two Combined Corps (41 &42)				
41 CC (2009)	**41 CC (2011)**	**41 CC (2013)**	**41 CC (2014)**	
• 1 Mech Div (RRU) • 1 Mot Div • 1 Armd Bde • 1 Arty Bde • 1 AD Bde • 1 Engr Regt	• 1 Mech Inf Div (RRU) • 1 Mot Inf Div • 1 Armd Bde • 1 Arty Bde • 1 AD Bde • 1 Engr Regt	• 1 Mech Inf Div (RRU) • 1 Mot Inf Div • 1 Armd Bde • 1 Arty Bde • 1 AD Bde • 1 Engr Regt	• 1 Armd Bde • 1 Mech Inf Div (RRU) • 1 Mot Inf Div • 1 Arty Bde • 1 AD Bde • 1 Engr Regt	• Offensive CC • All formations were brigadized in 2009 less the motorized infantry division and the mechanised infantry division (RRU).
42 CC	**42 CC**	**42 CC**	**42 CC**	
• 1 Mot Div • 1 Amph Div (RRU) • 1 Arty Div	• 1 Mot Inf Div • 1 Amph Div (RRU) • 1 Arty Div • 1 Armd Bde • 1 AD Bde • 1 Avn Regt	• 1 Mot Inf Div • 1 Amph Mech Div (RRU) • 1 Arty Div • 1 Armd Bde • 1 AD Bde • 1 Avn Regt	• 1 SF Unit • 1 Armd Bde • 1 Amph Mech Div (RRU) • 1 Mot Inf Div • 1 Avn Bde • 1 Arty Div • 1 AD Bde • 1 Engr Regt • 1 EW Regt	• Amphibious CC • A major reorganization of the formations on the anvil and needs to be monitored.

MR or MD Units (2009)	MR or MD Units (2011)	Other Forces (2013)	Other Forces (2014)	
• 1 EW Regt • 1 Inf Bde • 1 SAM Bde • 1 Avn Regt • 1 Pontoon Br Bde	• 1 Mot Inf Bde • 1 (Composite) Mot Inf Bde (Composed of units Drawn from across the PLA and deployed in Hong Kong on a rotational basis) • 1 AD Bde • 1 SSM Bde • 1 EW Regt • 1 SF Unit	• 1 Mot Inf Bde • 1 (Composite) Mot Inf Bde (Composed of units Drawn from across the PLA and deployed in Hong Kong on a rotational basis) • 1 AD Bde • 1 EW Regt • 1 SF Unit	• 1 Mot Inf Bde • 1 (Composite) Mot Inf Bde (Composed of units Drawn from across the PLA and deployed in Hong Kong on a rotational basis)	

Trends of Transformation Chengdu MAC Two Combined Corps (13 & 14)				
13 CC (2009)	**13 CC (2011)**	**13 CC (2013)**	**13 CC (2014)**	
• 2 Mot Div (149th Div RRU) • 1 Arty Div • 1 Armd Bde • 1 AD Bde • 1 Avn Regt • 1 Engr Regt	• 1 (High Alt) Mech Inf Div (RRU) • 1 Mot Inf Div • 1 Armd Bde • 1 Arty Bde • 1 AD Bde • 1 Avn Regt • 1 Engr Regt	• 1 (High Alt) Mech Inf Div (RRU) • 1 Mot Inf Div • 1 Armd Bde • 1 Arty Bde • 1 AD Bde • 1 Avn Bde • 1 Engr Regt	• 1 SF Unit • 1 Armd Bde • 1 (High Alt) Mech Inf Div (RRU) • 1 Mot Inf Div • 1 Arty Bde • 1 AD Bde • 1 Avn Bde • 1 Engr Regt • 1 EW Regt	• Dual Task CC • The formations have not been reorganized primarily because their areas of employment are in near vicinity. The formations of the group army have been regularly employed in quelling unrests in Tibet. • With the additional SF Unit and the EW Regt, the CC has an Orbat akin to an offensive CC.
14 CC	**14 CC**	**14 CC**	**14 CC**	
• 4 Mot Div • 1 Arty Div • 2 Armd Bde • 1 Arty Bde • 2 AD Bde • 1 Avn Regt	• 1 (Jungle) Mot Inf Div • 1 Mot Inf Division • 1 Armd Bde • 1 Arty Bde • 1 AD Bde	• 1 (Jungle) Mot Inf Div • 1 Mot Inf Division • 1 Armd Bde • 1 Arty Bde • 1 AD Bde	• 1 (Jungle) Mot Inf Div • 1 Mot Inf Division • 1 Armd Bde • 1 Arty Bde • 1 AD Bde • 1 Engr Regt	• Dual Tasked CC

MR or MD Units	MR or MD Units	Other Forces	Xixang MD	
• 2 Mtn Inf Bde (may be light or mech inf) • 1 EW Regt • 1 Avn Regt • 1 Spec Ops Unit	• 1 (High Alt) Indep Mech Inf Bde • 2 Indep Mtn Inf Bde • 1 EW Regt • 1 Spec Ops Unit	• 1 (High Alt) Mech Inf Bde • 2 Mtn Inf Bde • 1 EW Regt • 1 SF Unit	• 1 SF Unit • 1 (High Alt) Mech Inf Bde • 2 Mtn Inf Bde • 1 Arty Regt • 1 AD Regt • 1 Engr Regt • 1 EW Regt	• Arty, AD and Engr Regt added to the Orbat of the MD troops • The MD has now a similar composition as an offensive CC
Trends of Transformation Lanzhou MAC **Two Combined Corps (21 & 47)**				
21 CC (2009)	**21 CC (2011**	**21 CC (2013)**	**21 CC (2014)**	
• 1 Armd Div • 1 Mot Div (RRU) • 1 Arty Bde • 1 AD Bde • 1 Engr Regt	• 1 Armd Div • 1 Mot Inf Div (RRU) • 1 Arty Bde • 1 AD Bde • 1 Engr Regt	• 1 Armd Bde • 1 Mech Inf Bde • 1 Mot Inf Div (RRU) • 1 Arty Bde • 1 AD Bde • 1 Engr Regt	• 1 SF Unit • 1 Armd Bde • 1 Mech Inf Bde • 1 Mot Inf Div (RRU) • 1 Arty Bde • 1 AD Bde • 1 Engr Regt • 1 EW Regt	• Offensive CC • All formations are brigadized, less the motorized infantry division.
47 CC	**47 CC**	**47 CC**	**47 CC**	
• 1 Armd Bde • 1 Mech Bde • 2 Mot Bde • 1 Arty Bde • 1 AD Bde • 1 Engr Regt	• 1 Armd Bde • 1 Mech Inf Bde • 2 (High Alt) Mot Inf Bde • 1 Arty Bde • 1 AD Bde • 1 Engr Regt	• 1 Armd Bde • 1 Mech Inf Bde • 2 (High Alt) Mot Inf Bde • 1 Arty Bde • 1 AD Bde • 1 Engr Regt	• 1 Armd Bde • 1 Mech Inf Bde • 2 (High Alt) Mot Inf Bde • 1 Arty Bde • 1 AD Bde • 1 Engr Regt	• Dual Task CC • All formations were brigadized in 2009. In 2011 the motorized brigades were converted into terrain specific formations.

Xinjiang (2009)	Xinjiang (2011)	Xinjiang MD (2013)	Xinjiang MD (2014)	
• 2 Mot Div • 2 Mtn Div • 1 Arty Bde • 1 AD bde • 2 Inf Regt • 2 Indep Inf Regt • 1 Engr Regt	• 1 (High Alt) Mech Div • 3 (High Alt) Mot Div • 1 Arty Bde • 1 AD bde • 2 Indep Mech Inf Regt • 1 Engr Regt • 1 Avn Bde	• 1 (High Alt) Mech Div • 3 (High Alt) Mot Div • 1 Arty Bde • 1 AD bde • 2 Indep Mech Inf Regt • 1 Engr Regt • 1 Avn Bde	• 1 SF Unit • 1 (High Alt) Mech Div • 3 (High Alt) Mot Div • 1 Indep Mech Inf Regt • 1 Avn Bde • 1 Arty Bde • 1 AD bde • 1 Engr Regt • 1 EW Regt	• A number of formations need to be reorganized. In a major change, in 2011, the mechanised and motor-ised forma-tions were converted into ter-rain specific formations. However the formations of the MD have been regularly employed to quell upris-ings/ unrests in Tibet and Xinjiang. • SF Unit and EW Regt added to the Orbat of the MD troops • The MD has now a similar composition as an offen-sive CC

		Xi'an Flying Academy: 2 Trg Bde with CJ-6; JL-8; Y-7; Z-9		
MR or MD Units	**MR or MD Units**	**Other Forces**	**Other Forces**	
• 1 Engr Bde • 1 AAA Bde • 1 Spec Ops Unit	• 1 EW Regt • 1 Spec Ops Unit	• 1 EW Regt • 1 SF Unit		

The analysis of the transformation of PLAA in last five years reveals the following:

- The basic organization of a GA or a CC has almost been standardized to the fighting elements, which can be armoured brigade/ division, mechanized brigade/ division or motorized brigade or division supported by an arty brigade, AD brigade and an engineer regiment.

- The offensive CC now has a SF Unit and an EW Regiment in their Orbat.

- From 2013 to 2014, the maximum transformation has taken place in the Chengdu and Lanzhou MACs.

- In the Shenyang (40 CC), Beijing (27 CC), Jinan (20 CC & 54 CC) and Lanzhou (47 CC) MACs one group army is today completely brigadized. The remaining CCs in these MACs are left with one or two divisions which need to be reorganized.

- In many MACs, all formations less one mechanized/ motorized divisions have been reorganized into brigades. The mechanized / motorized divisions left are the designated RRUs. 54 CC, 31 CC and 21 CC are the examples of this.

- In the Nanjing MAC, the formations which are likely to be employed for the Taiwan contingency remain as divisions. However the remaining divisions have been reorganized into brigades. For example in the 1 CC, the amphibious mechanized infantry division and artillery divisions have not been reorganized, whereas the remaining formations are all brigades. Similar dispensation can be seen in the 41 and 42 CCs of the Guangzhou MAC.

- Maximum number of formations brigadized are in the Jinan MAC, there by supporting the fact that the formations of this MAC are likely to be employed initially in any contingency across China.

- In the Chengdu MAC, the 13 and 14 CCs have one armoured brigade each. The remaining formations are divisions, with some of them having terrain specific employment like high altitude and jungle. It can be assumed that this dispensation is likely to stay as they are close to their areas of employment and are likely to be employed in the initial phases should such a contingency arise.

- The CCs which need to be closely watched for reorganization are 38 & 65 CCs (Beijing MAC) and 41 & 42 CCs (Guangzhou MAC). In addition the following also should be monitored:

  ➢ The participation of the various formations in the trans MAC exercises (TRSO exercises).

  ➢ The detailed changes in the equipment profile of the formations.

Some identifications of formations and sub units. Details of formations and sub units identified in the Lanzhou and Chengdu MAC are as given in Appendix 1

Training

China's ground forces remain challenged by a lack of combat experience and self-identified limitations in the leadership abilities of its command staff, particularly at operational levels. These problems have long been exacerbated by a lack of realism in training. However, the PLA began executing plans in 2009 designed to help overcome these issues by 2020, including increased force-on-force training against dedicated opposing force units, adopting simulator use for training, developing automated command tools to aid command decisions, and increasing the education levels and science and technology training of PLA commanders and staff officers.

In 2009, the PLA organized a trans-MAC actual-troop exercise codenamed "Stride 2009" participated by the Shenyang MAC, the Lanzhou

MAC, the Jinan MAC, the Guangzhou MAC and the PLA Air Force, in which nearly 50,000 troop's participated and maneuvered for over 50,000-odd kilometers[12]. Unlike previous annual tactical exercises, the Army Divisions and their supporting Army Aviation units were deployed in unfamiliar areas far from their garrison training bases by civilian rail and air transportation assets. In 2010, a mechanized infantry division from the Shenyang MAC was transported northwest to the Lanzhou MAC. Troops from the Jinan MAC to the East and the Guangzhou MAC to the South were mobilized and deployed in the similar manner[13]. In these unprecedented exercises, one of the PLA's major objectives this time was to improve its capacity for long-range force projection. All heavy weapon systems, such as MBTs and AFVs, were carried by rail, and lightly armoured troops deployed to the Jinan Military Command by China Railway High-speed (CRH) trains travelling at speeds up to 350 kmph. When the Mechanized Infantry Division from the Lanzhou Military Command headed for the Taonan Tactical Training Base in Jilin Province of the Shenyang Military Command, more than 700 vehicles advanced across the Yellow River over a 250-metre long pontoon bridge built by the Lanzhou Division's engineering regiment at a ferry place in northwest Ningxia Hui Autonomous Region. The Division's forward command post elements took off from a northeastern air base on two civilian Boeing 737-800 airliners. Another group of heavy equipment carried on two chartered trains also departed from railway stations in Yinchuan and Qingtongxia. This was followed up by the rest of the 13,000 troops of the Lanzhou Division being transported through five provinces and regions, day and night, to reach the Taonan Tactical Training Base. The Lanzhou Military Command also deployed combat aircraft and attack helicopters to provide air cover for the long-distance manoeuvres, and also participated in the live-fire drills at the tactical training base of the Shenyang Military Command. On September 1, about 10,000 soldiers and 1,000 vehicles of a Division under the Shenyang Military Command were deployed at the foot of the Helan Mountain in Northwest China. The last participating components of Stride 2009 debuted on September 9, this being a Motorized Infantry Division of the Guangzhou MC that crossed the Xiangjiang River in a forced-march and pressed on toward Central China. In the following four days, the soldiers traversed four provinces and autonomous regions spanning 2,000 km[14].

In October 2010, the PLA organized another trans-MAC mobile exercise codenamed "Mission Action 2010". Some troops totaling 30,000 from three combined corps under the Beijing MAC, the Lanzhou MAC and the Chengdu MAC participated in the exercise[15]. In these exercises the PLA introduced a newly developed laser-beam combat simulation system through which the troops in each base were divided into 'Red' and 'Blue' rivals to carry out simulated battles. The troops involved in the exercises also adopted China's Beidou or COMPASS-G2 satellite communications and positioning system for encrypted communications between the PLA's headquarters and the Divisions to reduce the dependence on foreign space-based communications systems[16].

Another prominent exercise to be carried out by the PLA was the 20-day 'Airborne Manoeuvres 2009', the largest such exercise to be carried out to date by the PLA's 15th Airborne Army. Kicking off on October 18, it has been described as being a *'trans-theatre comprehensive campaign manoeuvre exercise'* involving 13,000 officers and men, 1,500-plus vehicles and 7,000-odd units of equipment. During the exercises, the participating troops manoeuvred more than 2,000 km between Hubei, Henan, Anhui and Jiangsu provinces with multiple means and routes. Besides, they also assembled at the designated places twice to carry out live-shell, multi-terrain and multi-subject exercises under close-to-actual-war conditions[17]. The 15th Airborne Army is the PLA's Strategic Rapid Reaction Unit and reports directly to the Central Military Commission. Its principal mission is to intervene rapidly into China's internal trouble spots as was done during the 1988 Tibetan riots when 10,000 airborne troops were airlifted in less than 48 hours. During the Sichuan earthquake in May 2008, 4000 airborne troops arrived within 36 hours for spearheading the disaster relief operations[18].

In September 2013, the PLAA carried out another trans-MAC campaign exercise codenamed "Mission Action 2013" in the 31st Combined Corps under the Nanjing Military Area Command (MAC) of the Chinese People's Liberation Army (PLA). The exercise was conducted in three echelons. The participating troops of 17,000-odd officers and men were mainly from related forces under the 31st Combined Corps. The East China Sea Fleet and the South China Sea Fleet of the PLA Navy and the air force under the Nanjing MAC dispatched surface ships, transport aircraft, fighters and armed helicopters to take part in the force projection and the actual-troop drills. Civil aviation and railway sectors also participated in long-range force projection[19].

PLAA and Force Projection

As Larry Wortzel says "Inside China and in its immediate neighbourhood, the PLAA is an effective force capable of fast mobilization, reinforced with local forces and reserves, and rapidly deployed in large numbers. However it is also the service that has been the slowest to modernize and exhibited the slowest speed of adopting information age electronic systems"[20]. As the PLAA is most likely to face armies in its immediate neighbourhood, the speed with which it can mobilize, its ability to mass forces and the sheer number of forces available implies that PLAA is the predominant land force in the region.

In terms of training, the PLAA is now conducting regular force on force training, a deviation from the past. This training improves the capacity of the leaders to think independently and react quickly to the changing battle field conditions. AS far as mobilization and movement is concerned PLAA has an excellent track record. The PLAA has the capacity to take control of and use for its support civilian rail, aircraft, road systems and transport for national security reasons during mobilization. Whenever the PLAA moves large forces by rail road or air, it diverts all civilian traffic. All of this can move PLAA rapidly inside China and is important for disaster management and mitigation, dealing with incidents of civilian unrest and positioning mass forces against an enemy on China's immediate periphery[21].

Today the PLAA's weakness is its capacity to project decisive forces over long distances away from its border. If PLAA continues to modernize and develop at the current pace, by 2025 it will be capable of conduction forced insertions in "the face of hostile forces at distances away from China on the scale of France, UK or even Russia"[22]. To match the capacity and capability of US, there is still a lot of work to be done by PLAA.

People's Armed Police Force (PAPF)

Article 22 of the PRC Law on National Defence adopted on March 14, 1997 states "The armed forces of the People's Republic of China are composed of the active and reserve units of the PLA, the Chinese People's Armed Police Force (PAP), and the people's militia[23]." The missions of this three tiered force are defined as:

"The active units of the Chinese People's Liberation Army are a standing army, which is mainly charged with the defensive fighting mission. The standing army, when necessary, may assist in maintaining public order in accordance with the law.

Reserve units shall take training according to regulations in peacetime, may assist in maintaining public order according to the law when necessary, and shall change to active units in wartime according to mobilization orders issued by the state. Under the leadership and command of the State Council and the Central Military Commission, the Chinese People's Armed Police force is charged by the state with the mission of safeguarding security and maintaining public order. Under the command of military organs, militia units shall perform combat-readiness duty, carry out defensive fighting tasks, and assist in maintaining the public order"

The **People's Armed Police**, officially **People's Armed Police Force** (PAPF), is a paramilitary or gendarmerie force primarily responsible for civilian policing and fire rescue duties in the People's Republic of China, as well as provide support to PLA during wartime. In contrast to public security police, PAP servicemen, also called as "Armed Policemen, wear olive green instead of the blue uniforms of the Public Security Department People's Police and other branches of People's Police. From January 1, 2005 to July 31, 2007 the position had been renamed 'internal guard' with arm insignia reflecting this change; new uniforms issued on August 1, 2007 carried to term for "People's Armed Police Force".

History

The history of the People's Armed Police is as long as that of the People's Republic, and its origin can be traced back to the People's Liberation Army, which was responsible for both defending the nation from foreign invasions and internal security. Although the force was officially established in 1982, its constituent units stretch back to 1949. After the establishment of the People's Republic of China, it was soon apparent that the different troops were required for the vastly different missions, and the domestic security functions had to be removed from the People's Liberation Army. As a result, the portion of People's Liberation Army responsible for internal security and other domestic police missions branched out to form the Public Security Army, under the administration of the Ministry of Public Security of the People's Republic of China. Although under the Ministry of Public Security, the Public Security Army troops were not exactly public security police officers because in addition to regular police work, they were also tasked with secondary military tasks which were not part of the responsibility of regular police officers of the public security ministry. After numerous name changes and reorganization, the PAP was created on June 19, 1982 by an amalgamation of the PLA's border control, internal security units (domestic 'internal guard' or and fire department, as well as from Ministry of Public Security units[24].

Mission

The PAP's primary mission is internal security. The first law on the People's Armed Police, the Law on the People's Armed Police Force (PAPF), was passed in August 2009, giving it statutory authority to respond to riots, terrorist attacks or other emergencies. Such units guard government buildings at all levels (including party and state organizations, foreign embassies and consulates), provide personal protection to senior government officials and provide security functions to public corporations and major public events. Some units perform guard duty in civilian prisons and provide executioners for the state. The PAP also maintains tactical counter-terrorism (CT) units in the Immediate Action Unit (IAU), Snow Wolf Commando Unit (SWCU) and various Special Police Units (SPU).

PAP border security forces guard China's land and sea borders, as well as its ports and airports. Other units guard China's forests, gold mines and hydropower facilities, as well as provide traffic-policing, Fire Fighting and road construction services. Among them, the Border Security and Fire Fighting are under the control by both Central Military Commission and Ministry of Public Security (which also controls the People's Police) while the other forces are under the control of CMC only. The border security force in particular, is also a law enforcement agency. The Coast Guard, part of the Border Control Department (BCD), is now under the State Oceanic Administration formed in 2013[25].

The secondary mission of the PAP is external defence, and in times of war PAP internal security units can act as light infantry supporting the PLA in local defence missions[26].

Command and Control

As per the China's National Defence in 2006, the defence white paper, PAPF is a component of China's armed forces and subordinate to the State

Council, the PAPF is under the dual leadership of the State Council and the CMC.

The State Council exercises leadership over the PAPF through relevant functional departments, assigns routine tasks to it, decides its size and number of organizations, and is responsible for its command, operations, and financial and material support. However the PAPF has an independent budgetary status in the financial expenditure of the state.

The CMC is responsible for the PAPF's organizational structure, management of officers, command, training and political work. It exercises leadership over the PAPF through the four general headquarters/departments. In terms of conducting public security operations and relevant capability building, the PAPF General Headquarters is under the leadership and command of the Ministry of Public Security and the PAPF units at and below the contingent level are under the leadership and command of the public security organs at the same level.

Strength and Organization

The PAP is estimated to have a total strength of 1.5 million, with over half its strength (800,000) employed in its internal security units. However, the Defence White Paper of 2006 and IISS Military Balance 2011 figures put the number at 660,000. These units are organized in division-sized elements and are located in each province, autonomous region and centrally-controlled city. Some provinces have more than one internal security *zongdui,* equal to a division size force of light infantry, due to the transfer of 14 PLA Divisions (numbering 500,000 personnel) to the PAP during the late 1990s. Additionally, the PAP maintains a national headquarters in Beijing.

Strength. As per the Military Balance 2011[27] published by the IISS the strength of the PAPF is 660,000. The details are as under:-

- Internal Security Forces – 14 Mobile Divisions and 22 Mobile Independent Regiments including Fire Fighting and Garrison Units.

- Border Defence Force.

  - ➢ 30 Division Headquarters, 110 Border Regiments and 20 Marine Regiments.

> ➢ Patrol and Coastal Combatants – 154+.

  - • Patrol Craft Offshore – 16.

  - • Patrol Boat/ Patrol Boat Fast – 138+.

Force Building

The PAPF is working to strengthen itself as part of the ongoing military Modernisation of the Chinese Defence Forces. It is enhancing staff competence, and conducting strict management so that its personnel can fully perform their duties. Using the national information infrastructure, the PAPF has established a preliminary system of three-level integrated information networks, linking general headquarters with the grass-roots squadrons. It has made progress in real-time command and control, management of duties through visual means, networked education and training, and office automation. The PAPF possesses a basically complete range of equipment through R&D and procurement of urgently needed weaponry and equipment. It has set up and improved a distinctive mechanism for the selection, training and employment of officers and NCOs. In particular, priority is given to the training of inter-disciplinary personnel. The PAPF conducts mission-oriented training on a priority basis to better perform guard duties, manage emergencies and combat terrorism.

The PAPF is steadily improving its logistical support system based on self-support and supplemented by social and PLA support to raise the efficiency of integrated support. It runs a crisis response support system covering the three echelons of the general headquarters, contingents (divisions) and detachments (regiments), to better respond to emergencies, and unexpected and complex situations. It promotes standardized and institutional logistical management by exploitation of IT and uniformly standardizes its facility configurations, work procedures, operating mechanisms and management requirements. The PAPF is pursuing reforms in housing, procurement of bulk materials and project procurement, medical care, and outsources food, barracks and bedding and clothing services.

In recent years, the PAPF has conducted friendly exchanges with the armed police forces, military police, internal security forces, public security forces and other similar forces of more than 30 countries to draw on each other's practices and cooperate in conducting anti-terrorism training. Its medical personnel, as part of Chinese rescue teams, have participated in disaster-relief missions in the aftermath of the earthquakes in Iran, Pakistan and Indonesia, and the tsunami in the Indian Ocean.

Certain Activities of PAPF in Addition to Its Primary Missions[28]

China's Constitution and laws stipulate that an important task for the country's armed forces is to take part in national construction, emergency rescue, and disaster relief. This was enunciated in detail in the China's National Defence in 2010 issued by the Information Office of the State Council in March 2011. According to the White Paper the activities of the People's Liberation Army (PLA) and the People's Armed Police Force (PAPF) in last two years are

- They have engaged a total of 1.845 million troop deployments in disaster relief operations.

- They have rescued or evacuated a total of 1.742 million people, rush-transported 303,000 tonnes of goods, dredged 3,742 km of waterways, dug 4,443 wells, and fortified 728 km of dikes and dams.

- The PLA and PAPF have actively participated in and supported national construction work, of which a key component is the large-scale development of the western region.

- They have participated in construction of more than 600 major infrastructure projects relating to transportation, hydropower, communications and energy.

- They have set up more than 3,500 contact points for rural poverty alleviation, and provided assistance to over 8,000 small public initiatives, such as water-saving irrigation projects, drinking water projects, road construction projects, and hydropower projects.

- PAPF medical and health units have provided assistance to 130 county-level hospitals in poverty-stricken western areas, sent there 351 medical teams.

Reserve Force

With active servicemen as its backbone and reserve officers and men as its foundation, the reserve force is an armed force formed in line with the unified structure and organization of the PLA. It is under the dual leadership of the PLA and local Party committees and governments. The positions of chief military and political leaders at all levels and principal department leaders, as well as a proportion of the staff members and professionals and specialists, are assumed by active servicemen. Reserve officers are chosen mainly from qualified retired servicemen, civil officials, cadres of the people's armed forces departments, cadres of the militia and civilian technicians with the appropriate military specialties. Reserve soldiers are chosen mainly from qualified discharged soldiers, trained primary militia members, and civilians with the appropriate military specialties. The Reserve Force was founded in 1983 and in August 1986 it formally became the part of the PLA. In May 1995 the NPC Standing Committee adopted the Law of the Peoples Republic of China on Reserve Officers. In April 1996, the CMC began to confer military ranks on reserve officers. The Law of the Peoples Republic of China on National Defence in March 1997 explicitly stipulates that China's Armed forces consists of the active duty force and the reserve force of the PLA, the People's Armed Police Force and the militia. IISS Military Balance 2011 gives the strength of the Reserves as 510,000[29], which is organized into:

- Armoured   – 2 Regiments.

- Infantry   – 18 Divisions, 4 Brigades, 3 Regiments.

- Artillery   – 3 Divisions; 7 Brigades.

- Air Defence   – 17 Divisions; 8 Brigades; 8 Regiments.

- Engineers   – 1 (Pontoon Bridging) Brigades; 15 Regiments; 3 (Pontoon Bridging) Regiments.

- Logistic   – 9 Brigades; 1 Regiment.

- Signals   – 10 regiments.

- Chemical   – 7 Regiments.

- There are no reserves mentioned for the PLAAF, PLAN and the Second Artillery.

In recent years, the reserve force has undergone consistent improvement in various aspects of its building and reform. It works to improve its organizational models on a regional basis, to explore a systematic and organic organizational model based on new and high-tech industries, and to develop such organizational models as personnel-and-equipment organization, trans-regional organization and community-based organization. Based on possible wartime assignments, the reserve force has revised and updated the guidelines for its military training and evaluation, strengthened integrated training with active PLA units, and conducted on-base, simulated and networked training. Reserve officers and men are required to devote 240 hours to political education and military training each year. To be able to respond to emergencies in peacetime and to fight in war, the focus of the reserve force is shifting from quantity and scale to quality and efficiency, from a combat role to a support role, and from the provision of general-purpose soldiers to soldiers with special skills. It is working to become an efficient auxiliary to the active force and a strong component of the national defence reserve.

Details of the PAPF units and Reserve's in the Lanzhou and Chengdu MAC are attached as Appendix 2.

PLA Militia

The militia force is an important component of China's armed forces as well as the backup force of the PLA. In recent years, through transformation and reform, it has made progress in restructuring, in training reform, and in equipment building. China now has 8 million primary militia members[30].

The militia force gives priority to reinforcing those units which are tasked with defending border and coastal areas, providing service support for different arms and services, and responding in emergencies. It has been realigned to extend from rural to urban areas as well as to areas along important communication lines, from ordinary locations to key sites and areas, and from traditional industries to new and high-tech ones. As a result, its structure and layout have been further improved. In line with the newly revised Outline for Military Training and Evaluation of the Militia, it promotes reforms in military training, holds joint training and exercises with active PLA units, improves the construction of associated training base facilities at all levels, and attaches importance to key detachment training. Its capabilities in dealing with both emergencies and wars have been greatly enhanced. The militia strengthens its building of equipment for the purposes of air defence, emergency response, and maintaining stability, supply of new types of air defence weaponry and equipment, and retrofitting and upgrading of existing weapons. There have been significant increases in the level of equipment-readiness and in the full kit rate (FKR).

The militia has taken an active part in such operations as counter-terrorism, stability maintenance, emergency rescue, disaster relief, border protection and control, and joint defence of public security, and has played a unique role in accomplishing diversified military tasks. Each year, it mobilizes more than 90,000 militiamen to serve as guards on bridges, tunnels and railways, more than 200,000 to take part in joint military-police-civilian defence patrols, more than 900,000 to participate in emergency response, rescue and relief operations following major natural disasters, and nearly 2 million to engage in the comprehensive control and management of social order in rural and urban areas.

Trends to be watched in times ahead

To correctly assess the Modernisation trajectory of the PLAA the following major trends need to be continuously watched and analysed:

- Weather the PLA will increase in size of further downsize, will depend largely on internal dynamics and the geo – politics of the world. However it can be predicted that the percentages of naval and air forces will increase as ground forces decrease. Moreover the Local War segment of the PLA will continue to grow and the Peoples War segment will shrink in size.

- PLA civilians and business operations may be stripped from the active duty rolls.

- The numbers of reserves and People's Armed Police will increase.

- For the foreseeable future, units will have a mix of high, medium and low technology weapons and equipment and will strive to find ways to maximize the use of their existing equipment to defeat a high-technology enemy.

- The numbers and types of logistics and technical units will increase throughout the force to maintain and support the PLA's modern equipment.

- The Chinese defence industries will be able to produce limited amounts of modern weapons for the PLA, but most truly advanced weapons will be of foreign origin and relatively few in number.

- Rapid deployment of conventional forces will be enhanced through acquisition of transport ships and aircraft as well as by unit consolidation near points of embarkation.

- Changes in the command and control structure will contribute to better integration of forces and capabilities.

- Several regional headquarters will be eliminated, resulting in "theater like" headquarters.

- A few smaller headquarters will be formed for the Army, Special Operations Forces, and Space Forces.

- Tactical units will be restructured during a period of experimentation.

- As far as India is concerned, the force levels along the LAC India will remain unchanged. However Indian Army will contest a more capable PLA, with hi tech electronic support and a large asymmetric capability.

PLAA in 2025

PLAA is the largest component of the PLA and will continue to be so even till 2025. Other changes that one may observe in the PLAA are elucidated in the succeeding paragraphs;

- The PLAA will continue to modernize its ground force formations for offensive operations along five axis in its immediate priphery: Korea/Japan (Japan a higher priority); Taiwan; Southeast Asia; India; Russia.

- The size of the PLAA is likely to remain same or downsize and this will depend on the overall downsizing trends of the PLA. With increased brigadization and achieving higher levels of informationalization, there will be an overall decrease in the strength of the PLAA. Within the PLA where there is a likely increase anticipated is the number of logistics and technical units. Though technically not under the PLAA, there will be an increase in the strength of the PAPF, primarily due to the restive periphery and as an offset of the downsizing. The strength of the Reserves is also likely to increase to cater for various contingencies and maintain cap on the strength of the PLAA.

- The PLAA will remain a regional force with limited force projection capability over long distances away from its borders.

- The key focus area of the PLAA will be rapid deployment and this capability will continue to increase as PLAA exercises and learns.

- In the PLA units and formations following trends may be seen

  ➢ The PLAA Units will be equipped with a mix of high, medium & low tech weapons and equipment. The endeavor of the PLAA will be to maximize the use of this mix to achieve the desired results.

  ➢ By 2025 almost 50 to 60 per cent of PLAA units will be equipped with cutting edge hi tech weapons systems and equipment

- By 2025 in the PLAA there will be a greater integration of forces & their synergized application for a variety of tasks and contingencies.

The PLAA of 2014 looks quite different from its predecessor in the mid-1990s/early 2000s. In 2025, it will be different still - significantly smaller, more mobile and combat capable. However, the PLAA's improvement in capabilities in absolute terms (as measured against itself) is only half of the equation for victory. Future PLAA capabilities must also be measured relative to the capabilities of potential opponents and the likelihood that they, too, will continue to modernize and improve their own capabilities. No matter how much the PLAA 'mechanizes' and 'informationalizes,' future combat will not be any easier than in its light infantry days. More depends on the intellectual capabilities and levels of training of the PLAA's officers and NCOs to plan for and execute a new and unproven joint doctrine, the battlefield techniques and procedures they have developed through realistic training, and their ability to adapt to changing circumstances, rather than the capabilities of any single weapon or weapons system the force may acquire in coming years.

Endnotes

1   Shambaugh David, "Modernizing China's Military", California, 2002, p. 154.

2   Dennis J Blasko, "A New PLA Force Structure", PLA in Info Age, Rand Corporation, pp 260-261.

3   "The Diversified Employment of China's Armed Forces", Information Office of the State Council, Peoples Republic of China, April 2013

4   The International Institute for Strategic Studies, The Military Balance 2011, Routledge, London.

5   The International Institute for Strategic Studies, The Military Balance 2014, Routledge, London.

6   Il-76 Production, GlobalSecurity.org, accessed from http://www. globalsecurity.org/military/world/russia/il-76-production.htm on 12 September 2013.

7   Prasun K. Sengupta, Mighty Spectacle: China's National Day parade featured undisclosed technologies, Force, November 2009, p. 63, accessed from http://www.forceindia.net/Mighty per cent20Spectacle-pdf/Prasun per cent20Article.pdf on August 12, 2014.

8   Gaurav Sharma, People's Liberation Army Ground Forces Modernisation-An Assessment, Scholar Warrior, Spring 2010, p. 45, accessed from http://www. claws.in/images/journals_doc/SW per cent20J.62-84.pdf on May 13, 2014.

9   Ibid.

10 Larry M. Wortzel, The Dragon Extends its Reach, Potomac Books, Virginia, US, 2013, p. 85.

11 The International Institute for Strategic Studies, The Military Balance 2009, 2011, 2013 and 2014, Routledge, London.

12 Dong Zhaohui, 'Mission Action 2013' trans-MAC campaign exercise kicks off, China Military Online, September 11, 2013, accessed from http:// eng.chinamil.com.cn/news-channels/china-military-news/2013-09/11/ content_5483115.htm on August 4, 2014.

13 Prasun K. Sengupta, Mighty Spectacle: China's National Day parade featured undisclosed technologies, Force, November 2009, p. 64, accessed from http://www.forceindia.net/Mighty per cent20Spectacle-pdf/Prasun per cent20Article.pdf on September 16, 2014.

14 Ibid.

15 Dong Zhaohui, 'Mission Action 2013' trans-MAC campaign exercise kicks off, China Military Online, September 11, 2013, accessed from http:// eng.chinamil.com.cn/news-channels/china-military-news/2013-09/11/ content_5483115.htm on August 4, 2014.

16 Prasun K. Sengupta, Mighty Spectacle: China's National Day parade featured undisclosed technologies, Force, November 2009, p. 64, accessed from http://www.forceindia.net/Mighty per cent20Spectacle-pdf/Prasun per cent20Article.pdf on September 16, 2014.

17 Ibid.

18 Ibid.

19 Dong Zhaohui, 'Mission Action 2013' trans-MAC campaign exercise kicks

off, China Military Online, September 11, 2013, accessed from http://eng.chinamil.com.cn/news-channels/china-military-news/2013-09/11/content_5483115.htm on August 4, 2014.

20 Larry M. Wortzel, The Dragon Extends its Reach, Potomac Books, Virginia, US, 2013, p. 97.

21 Ibid.

22 Ibid.

23 Accessed from http://www.npc.gov.cn/englishnpc/Law/2007-12/11/content_1383547.htm on August 22, 2013.

24 People's Armed Police, fas.org, accessed from http://fas.org/irp/world/china/mps/pap.htm on August 22, 2013.

25 Andrew Erickson and Gabe Collins, New Fleet on the Block: China's Coast Guard Comes Together, China Real Time, March 11, 2013, accessed from http://blogs.wsj.com/chinarealtime/2013/03/11/new-fleet-on-the-block-chinas-coast-guard-comes-together/ on September 23, 2013.

26 Accessed from http://www.npc.gov.cn/englishnpc/Law/2007-12/11/content_1383547.htm on August 22, 2013.

27 The Military Balance 2011, International Institute of Strategic Studies, Routledge, London, 2011, p. 236.

28 "The Diversified Employment of China's Armed Forces", Information Office of the State Council, Peoples Republic of China, April 2013.

29 Military Balance 2011, International Institute of Strategic Studies, Routledge, London, 2011, p. 231.

30 The Militia, GlobalSecurity.org, accessed from http://www.globalsecurity.org/military/world/china/pla-militia.htm on December 23, 2013.

3

PLA Air Force (PLAAF) Modernisation

"To conquer the command of the air means victory; to be beaten in the air means defeat and acceptance of whatever terms the enemy may be pleased to impose..."

General Giulio Douhet, The Command of the Air, 1921

"For good or for ill, air mastery is today the supreme expression of military power and fleets and armies, however vital and important, must accept a subordinate rank."

Sir Winston Leonard Spencer Churchill, 1874 - 1965

Introduction

The PLAAF is the largest military air force in the world with over 6,000 military airplanes and over 398,000 active personnel[1] and today the Modernisation of the PLAAF is progressing at a steady pace. The goal of PLAAF Modernisation is to improve capability to conduct offensive and defensive operations such as strike, air and missile defence, power projection, and early warning and reconnaissance. The key areas of emphasis include increased introduction of 4th and 4.5 generation multirole aircraft and the new upgraded H-6K bomber to increase PLAAF strike capabilities, as well as developing the 5th generation fighters. To meet reconnaissance and early warning goals the PLAAF has fielded several Airborne Early Warning (AEW), Electronic Warfare, and command-and-control (C2) systems, such as the KJ2000 and have fielded several Unmanned Air Vehicles (UAVs). To address their power projection deficiency, they have purchased a limited number of used IL-76 from Russian and are developing the Y-20 heavy lift transport aircraft, and the Y-9 medium-lift transport. To clearly asses the trends of Modernisation of the PLAAF, one should not take a "symmetric" view, in which there is a direct comparison the development to any other modernizing air forces of the world. The Chinese are not trying to match the modern air forces like the US Air force system vs. system, but are pursuing more of a system-of-systems approach that exploits what they perceive to be adversary weaknesses or exploitable vulnerabilities.

History of PLA Air Force (PLAAF)

The PLAAF was founded on November 11, 1949 after Mao Zedong's communists defeated the nationalist forces of Chiang Kai-shek, by an announcement by the CMC stating that "the Air force has become the official branch of the PLA and the PLAAF begins its glorious history starting from nothing, from weakness to strength"[2]. The initial assistance in training and establishing the air force was provided by the Soviet Union as they sent their trainers and airplanes to China. But they had hardly started training when the Korean War started and the PLAAF was no match for the UN Air force, which primarily were the US Air Force and the US Navy. It was the UN Air force that controlled the skies over Korea. Drawing lessons from this, after the war, the PLAAF built seven aviation schools and named them the First through Seventh People's Liberation Army Air Force Academies and in four years trained nearly 6000 pilots and 24,000 maintenance crews[3]. The further development of PLAAF suffered a setback by the Sino-Soviet split in the 1960s after Mao and Soviet Premier Nikita Khrushchev disagreed over prospects for peaceful coexistence with the West, as well as over issues within the Communist bloc. The Soviets withdrew their aviation advisors and military aid. Then, in 1966, Mao's Cultural Revolution tore China apart. The PLAAF, like almost every other institution in China, was crippled by radical ideologues. Training was reduced to 12 months from 28 months, flying hours plummeted to 24 hours a year[4], and training manuals were destroyed—being labeled as capitalistic.

After the death of Mao, Deng Xiaoping set China on its course of economic development and military buildup, with economics taking priority. In 1991, PLAAF was 'shocked and awed' by US airpower during Desert Storm in Iraq. Lt Gen Liu Yazhou belittled Iraq's line of defence in the desert, during the first Gulf war as the last line of defence, which proved helpless against US airpower. He likened it to the Great Wall of China, a source of national pride but unable to stop invasions from the North.[5] Consequently, the PLAAF shed lingering Russian influence and learned from USAF. PLAAF leaders aspired to free their service from supporting ground forces to become a strategic force. Liu, the main advocate of strategic airpower, was seen as a disciple of Giulio Douhet, the Italian pioneer in airpower doctrine and strategy.[6]

Today the PLAAF is in the midst of a transition between the limited, largely air defence force it used to be, and a more advanced force with modern equipment, doctrine, and capabilities. While the size of the force has shrunk considerably over the past 15 years, the same could not be said about its quality and capabilities. With the introduction of new generation aircraft, precision-guided munitions, support aircraft that serve as force multipliers, and modernized C4ISR capabilities, the PLAAF is now not only capable of carrying out traditional missions such as air defence and support for ground forces, but could also undertake offensive strikes against ground and naval targets beyond China's borders.

Core Missions of the PLAAF

From the Core Missions of the PLA as given in Chapter One, the following three core missions for the PLAAF can be derived, which are as under:

- The first core mission is to defend China's airspace—particularly Beijing, headquarters of the Communist Party and the seat of government and is priority one. In the White Paper, The Diversified employment of China's Armed Forces, this aspect finds a mention as a major tasking of the command organs responsible for air defence at all levels. The Shenyang Military District, in northeastern China, bordering Russia, the Sea of Japan, and North Korea, is a close second priority. It features a layered defence of airpower, conventional missiles, anti-aircraft artillery, and early warning systems, also intended to help protect Beijing.[7]

- The second mission is preparation for an assault on Taiwan, the self-governing island the Nationalists fled to in 1949 and over which Beijing claims sovereignty. This task is assigned to the Nanjing Military District across the Taiwan Strait in the province of Fujian. Beijing has never relinquished the threat to use military force against Taiwan, especially if the government in Taipei declared independence and asked other nations to recognize it as the legitimate government of a sovereign Taiwan. In the event of a war with Taiwan, the PLAAF would be tasked to gain air superiority over the island and strait and cover an amphibious landing force. The PLAAF might also mount an airborne invasion into Taiwan, as the PLA's paratroopers are part of the PLAAF, not the Army.

- The PLAAF's third, and newest, core mission is to acquire the capability to project power into the South China Sea and the Pacific Ocean up to the Second Island chain. This island chain runs through Andersen Air Force Base on Guam to Japan, where USAF has bases at Kadena, Yokota, and Misawa.

The Trends of PLAAF Modernisation

The Defence White Paper "The Diversified Employment of China's Armed Forces" released in April 2013 lays down the responsibilities, themes and the direction of Modernisation of the PLAAF. The responsibilities of the PLAAF are concerned it states *"The PLA Air Force (PLAAF) is China's mainstay for air operations, responsible for its territorial air security and maintaining a stable air defence posture nationwide. It is primarily composed of aviation, ground air defence, radar, airborne and electronic countermeasures (ECM) arms."*[8] The direction in which the PLAAF will modernize has also been elucidated and it states *"In line with the strategic requirements of conducting both offensive and defensive operations, the PLAAF is strengthening the development of a combat force structure that focuses on reconnaissance and early warning, air strike, air and missile defence, and strategic projection. It is developing such advanced weaponry and equipment as new-generation fighters and new-type ground-to-air missiles and radar systems, improving its early warning, command and communications networks, and raising its strategic early warning, strategic deterrence and long-distance air strike capabilities."*[9] This direction of the PLAAF Modernisation has also been commented upon in the Annual Report to the Congress, Military and Security Developments involving People's Republic of China and it states *"China continues to field increasingly modern 4th generation aircraft, but the force still consists mostly of older 2nd and 3rd generation aircraft, or upgraded variants of those aircraft. Within two years of the J-20 stealth fighter's first flight in January 2011, China tested a second next generation fighter prototype. The prototype, referred to as the J-31, is similar in size to a U.S. F-35 fighter and appears to incorporate design characteristics similar to the J-20. It conducted its first flight on October 31, 2012. China continues upgrading its H-6 bomber fleet (originally adapted from the late 1950s Soviet Tu-16 design) with a new variant that possesses greater range and will be armed with a long-range cruise missile. China also uses a modified version of the H-6 aircraft to conduct aerial refueling operations for many of its indigenous aircraft, increasing their combat range. The PLA Air Force possesses one of the largest forces of advanced SAM systems in the world, consisting of a combination of Russian-sourced SA-20 battalions and domestically produced HQ-9 battalions. China's aviation industry is developing a large transport aircraft (likely referred to as the Y-20) to supplement China's small fleet of strategic airlift assets, which currently consists of a limited number of Russian-made IL-76 aircraft. These heavy lift transports are needed to support airborne command and control (C2), logistics, para drop, aerial refueling, and reconnaissance operations, as well as humanitarian assistance and disaster relief missions. Developments in China's commercial and military aviation industry indicate improved aircraft manufacturing, associated technology, and systems development capabilities. Some of these advances have been made possible by business partnerships with Western aviation and aerospace firms (including cleared U.S. defence contractors), which provide overall benefit to China's military aerospace industry. China will continue to seek advancement in aerospace technology, capability, and proficiency to rival Western capabilities."*[10]

RAND Corp in its broad assessment of the PLAAF, in a report titled *Shaking the Heavens and Splitting the Earth* has said *"China's air force is in the midst of a transformation. A decade ago, it was an antiquated service equipped almost exclusively with weapons based on 1950s-era Soviet designs."* Those weapons, RAND noted, were *"operated by personnel with questionable training according to outdated employment concepts. Today, the [PLAAF] appears to be on its way to becoming a modern, highly capable air force for the 21st century"*.[11]

Similarly, the National Air & Space Intelligence Center at Wright-Patterson AFB, Ohio, recently published a wide-ranging analysis of the PLAAF. According to their analysis *"The People's Liberation Army Air Force is an organization undergoing a series of major transitions and significant changes"*, the centre noted. *"Like the rest of the Chinese armed forces, change in the PLAAF is happening across a wide front, and in myriad endeavours: in operational matters, in institutional affairs, and in the acquisition of new capabilities. Today, the PLAAF is more operationally capable than at any time in its past, and it is enjoying the fruits of years of focused and sustained reform and Modernisation"*, the assessment continued, before cautioning that the PLAAF is coping with the *"inevitable dislocations and uncertainties attendant to institutional change on a grand scale. China's Air Force thus "remains challenged in many areas"*.[12]

From the above the following trends of PLAAF Modernisation clearly emerge;

- PLAAF must be capable of both offensive and defensive operations.

- The focus of future development will be on reconnaissance and early warning, air strike, air and missile defence, and strategic force projection.

- This trend of development will lead to the PLAAF having the next generation fighters i.e. stealth, an integrated air and space defence system with early warning and command and communications (C2) networks and strategic force projection capability.

- This will further ensure PLAAF has an enhanced strategic early warning, strategic deterrence and long-distance air strike capabilities by 2025 if not earlier.

- If this is the likely trajectory of PLAAF, then the force structure of the PLAAF by 2025 can be analysed by following the trends as given under:

  - ➢ Stealth and counter stealth.

  - ➢ Integrated air and space defence.

  - ➢ Force multipliers.

  - ➢ Strategic Projection Capability: development of multipurpose strategic airlifter and Modernisation of the 15th Airborne (AB) Corps.

  - ➢ Air Defence.

Stealth and Counter Stealth

Today the PLAAF is focusing on two aspects of defending its air space, firstly the PLAAF is gravitating towards "large-area defence" and moving away from "key point defence". Given the PLAAF's previous focus on defending cities, industry, and military bases, this is perhaps the biggest change in Chinese thinking. The forward edge of the battle must be pushed toward the enemy, and intercepts must occur earlier. The second aspect, in order to fight a superior enemy, PLAAF must be able to achieve "local superiority" i.e. plug holes in the air defence, allow forces to concentrate and create favourable conditions for Chinese forces to destroy their enemy and second is to develop the ability to "shoot and scoot", which can improve chances of survival. Development of stealth and counter stealth in an informationalized environment is a very important capability that the PLAAF is in the process of mastering.

In January 2011 China unveiled its fifth generation fighter, the J-20 which represents a significant step in the evolution of the Chinese aerospace industry. This fighter undertook its first test flight, when US Secretary of Defence, Robert M. Gates was in Beijing and it was done with a drum roll of publicity that the Secretary of Defence and his delegation could not miss[13]. This new aircraft displays stealth features and indicates a determination on China's part to shape new military capabilities in the period ahead. However there are certain issues which need a detailed clarity and which the PLAAF designers and engineers need to overcome are;

- The true capabilities of the radar absorbing coatings, sensors and stealth attributes are not known.

- There are issues of capabilities of material selection, quality control and ensuring structural strength which need to be addressed.

- It is also not known that if the sensor suite is capable of supporting stealth operations.

- Lastly, the engine at present is a Russian RD-93 engine, which have been exported to China for manufacturing the JF-17 fighter jets. The attempts to develop a jet engine for the J-20 have been unsuccessful.

As regards the development of the counter stealth capability, Wayne A. Ulman, a senior China analyst at the U.S. Air Force (USAF) National Air and Space Intelligence Center (NASIC) said in a testimony to the congressionally mandated U.S.-China Economic and Security Review Commission in May 2010[14] *"They are developing a network-centric kill chain to fuse data from an extensive and diverse sensor network. They are also working to reduce the signature of current aircraft designs and on developing a low-observable fighter. As the PLAAF gains access to reduced signature systems, it will allow the development of tactics, training, and procedures for use against low-observable threat systems."* He added that China has spent more than a decade working on technologies, systems, and procedures to detect, track, and engage stealth aircraft and cruise missiles.

China, having generated a certain quantum of expertise in this field, including learning from the designers, technicians and scientists imported from CIS countries where they had been rendered unemployed post the break-up of the Soviet Union, China invested significantly in the aerospace sector and the benefits are visible now. The progress has been much faster than predicted by western analysts. The phenomenal growth in its economy permits China increased investments in innovation and the result would be that by 2025 or so China will become the world's most important centre for innovation, overtaking the US and Japan.

Integrated Air and Space defence

The concept of PLAAF as a strategic air force was codified during the 17th National Congress of the Communist Party of China held in November 2007. The party congress directed the PLAAF to aim "to build a modernized strategic air force that will be compatible with the international stature of our country and capable of carrying out the historical mission of our armed forces."[15] Hence to be a strategic air force the PLAAF was required to participate in joint operations as well as independent strategic actions to support the military and national development strategy of the country.[16] The PLAAF intends to carry out this strategic mission through the use of "integrated air and space operations." Integrated air and space operations refer to the organic combining of airpower and space power to form an integrated air-space force. According to PLAAF analysts, the air and space battlefield is the main domain for information collection in which the space component plays an important role[17]. According to The Military Balance 2014, China has a total of 59 operational satellites[18]; the details of these are given in the table below:

Table 3.1: Details of Chinese Military Satellites

Ser No	Nomenclature	Numbers	Remarks
Communication		5	
1.	Zhongxing	5	Dual use telecom satellites for civil and military communications

Ser No	Nomenclature	Numbers	Remarks
Navigation/ Positioning/ Timing		17	
2.	Beidou-1	2	
3.	Beidou-2(M)	5	
4.	Beidou-2(G)	5	
5.	Beidou-2(IGSO)	5	
Intelligence, Surveillance & Reconnaissance (ISR)		25	
6.	Haiyang-2A	1	
7.	Yaogan Weixing	22	Remote Sensing
8.	Zhangguo Ziyuan	2	- Do -
ELINT/ SIGINT		12	
9.	Shijian-6	8	Four pairs – reported ELINT/ SIGINT role
10.	Shijian-11	4	Reported ELINT/ SIGINT role
	Total	59	

The final aim of the PLAAF's use of space is to build a network-centric force in which disparate forces divided by function and distance will be fused into an organic whole through informationalization or the use of information technologies. Networked capabilities will allow the PLAAF to carry out four activities: information, air, and space superiority; precision strike; rapid manoeuvre; and multidimensional support. These capabilities are intended to achieve information superiority across all domains. In fact, the level of network capabilities is said to define the level of Modernisation of the PLAAF by 2025. The capabilities derived from a space-enabled, networked air and space force will also better integrate disparate services into a joint force, an essential prerequisite for winning local wars under conditions of informationalization. PLAAF aims to realize jointness in two ways. First, space-enabled air operations allow the PLAAF to provide better operational support to other services, for example, through precision strikes. Second, the C4ISR (command, control, communications, computers, intelligence, surveillance, and reconnaissance) capabilities provided by satellites will allow all services to share a common battlefield picture and to better communicate with each other. Through the use of these capabilities, PLAAF practicing Air and Space Integrated Operations will be able to achieve synergies in which the whole is more than the sum of its parts[19].

Force Multipliers

An important trend of PLAAF's Modernisation efforts in recent years is the development and deployment of support aircraft serving as force multipliers to enhance the effectiveness of its combat aircraft. These support aircraft include tankers, AEW and AWACS aircraft, electronic warfare and intelligence collection aircraft, search & rescue aircraft etc. For example, during the rescue and disaster relief mission in the aftermath of the 2008 Sichuan earthquake, a PLAAF communication relay aircraft cruised over the disaster zone to provide command and control support for the rescue troops on the ground, demonstrating the sort of capabilities not possessed by the force only few years before. Towards this the PLAAF is developing special mission aircraft including the KJ-2000 AWACS based on the Il-76 platform. The Y-8 transport planes are being modified to undertake a variety of roles of Airborne Battlefield Command, AEW and intelligence gathering.

Strategic Force Projection Capability

The PLAAF has been seeking to acquire more long-range transport aircraft to improve its airlift capabilities. Additional to the 14 IL-76MD transport aircraft in service since the 1990s, China ordered another 30 IL-76s plus 4 IL-78 tankers in 2005, but the delivery of these aircraft has been postponed due to a production problem in Russia[20]. The PLAAF also has modified the H-6 bomber for AAR role. Today two variants, H-6U for PLAAF and H-6DU for PLAN are in service, however their capabilities are limited by the number and type of aircraft it can refuel and its operating range. At the same time, China has developed an indigenous four-engine turbofan transport aircraft[21], Y-20 similar to American C-17 and European A400M to replace its IL 76 transport aircrafts. The Y-20, though resembles contemporary Western aircrafts, it is inferior to them in many ways. The biggest problem area is the engines which is grossly under powered. Once mated with a better engine, the Y-20 could become the basis for a new generation of support planes including refueling tankers, AEW and AWACS aircraft, electronic warfare and intelligence collection aircraft and SAR aircraft. Once these aircraft are in place, China's not only power projection capabilities to regions beyond its border would be greatly enhanced but also greatly enhance the PLAAF capabilities for humanitarian assistance and disaster relief operations.

The 15th AB Corps of the PLAAF has been elevated to the level of Strategic Forces by the PLA. In times ahead the 15th AB Corps will be the arm of PLAAF which will be capable of long range rapid mobile operations. A glimpse of the capabilities that the 15th AB Corps intends to achieve was exhibited in the September 2008 exercise carried out in the training ground in Inner Mongolia Autonomous Region, in which Integrated Parachuting of paratroopers and heavy equipment was executed that too in front of 113 military observers from 36 countries[22].

Air Defence

PLAAF has always laid special emphasis on air defence. It has thus built a formidable air defence which is difficult to penetrate. Important points in this regard are covered below:-

- **Radar Coverage.** PLA has extensive network of radars throughout the country and most of these radars are of Russian origin or their upgraded and modernized indigenous versions. Amongst the radars, there are at least three over-the-horizon radars, which give the PLAAF the ability to detect aircraft and missiles (cruise and ballistic) at extended ranges. China is also working on counter-stealth and counter-network radar technology.

- **Surface to Air Missiles.** According to one estimate, China today has 600 surface-to-air missile launchers and 16,000 anti-aircraft guns for air defence[23]. The PLAAF's Surface-to-Air Missile Corps has been operating the S-300 (SA-10 Grumble) family of surface-to-air missile system since the mid-1990s. The S-300 missile system was regarded as one of the world's most effective all-altitude regional air defence system, comparable in performance to the US MIM-104 Patriot system. The number of systems has been increasing with the passage of time and in 2014, the PLAAF is operating a total of 160 S-300 launchers (32 S-300PMUs, 64 S-300PMU1s, and 64 S-300PMU2s) grouped into 10 SAM battalions (40 batteries) – a few more systems have been inducted since then[24]. It is reported that the PLAAF has over 1,000 of these missiles. The acquisition of the S-300 system has significantly improved the PLA's capability of denying Chinese airspace to enemy air forces. In particular, the latest S-300PMU2 system gives the PLAAF limited ballistic missile defence capability for the first time. The Chinese are reported to have their version of the missile called Hongqi-15 (HQ-15) – it is reported that only the missiles have been produced while rest of the systems continue to be imported. At least one such system (S-300PMU) has been deployed in Tibet since 2010. Chinese request for license production of the system was turned down by Russia. The AD capability will further be enhanced when the recently contracted S-400 SAM systems are operationalized in the PLAAF. It will be seen that the available SAM systems can provide protection to only a number of cities and vital points. It is for this reason that some of the obsolescent systems (including SAM-2s) are still in use in some locations. On the other hand, China displayed a number of the latest state-of-the-art missile and anti-aircraft gun system at Zhuhai Air show China in November, 2012. These are likely to replace some of the obsolescent systems in due course. The PLA and PLN have their own air defence network. The number of SAM launchers and AA guns mentioned earlier are inclusive of their assets as well.

- **Passive Air Defence Measures.**   Chinese literature on air defence lays special emphasis on passive air defence measures like dispersal of assets, camouflage and concealment and hardened shelters (including surplus ones) for aircraft and equipment. Though no hard intelligence is available on the subject but one should expect a large number of reinforced concrete blast pens on any airfield with many of them without any aircraft or with just dummy aircraft; the attacking pilot would thus be wasting their effort by attacking them.

Three trends in air defence Modernisation are clearly visible as PLAAF is modernizing. First is a shift from a "protective air defence" towards an "offensive air defence" driven by more effective offensive operations. This calls for reliance on "integrated attack and defence" in which the offense mounts more attacks on enemy airfields. Today a common thought among the PLAAF commanders is to "actively organize counterattack operations of various scales to distract and attrit the enemy, disrupt his plans, destroy his offensive posture, gradually move the enemy into a reactive mode, and, ultimately, seize the operational initiative"[25]. The second trend is toward an "information air defence." Information has become a core component of strength, and information superiority must be incorporated into the entire course of an air defence campaign. The third trend is movement of PLAAF towards a joint air defence, which is part of a larger trend in PLA thinking toward joint operations, something many in US military and US think tanks have long argued will cause the PLA great difficulty[26]. The entire force structuring and Modernisation will follow these trends towards 2025.

Current Inventory of the PLAAF

Currently the PLAAF and PLA Naval Aviation operate a mixed fighter fleet consisting of both older aircraft fielded in the 1980s, and newer designs introduced in the 1990s and later. The thousands of J-6 (MiG-19 Farmer) fighters that once made up the fighter fleet have been retired. The current inventory is composed primarily of third and fourth-generation fighters and fighter-bombers, including 800 to 1,000 J-7 (MiG-21 Fishbed) and J-8II fighters, 76 Russian-built Su-27 fighters, 95 to 116 Chinese-assembled J-11 fighters, 76 Russian Su-30MKK multirole fighters, and some 60 to 80 Chinese indigenous J-10 multirole fighters. Thus having approximately 28 per cent of its combat aircraft fleet comprising of aircrafts which can be classified as modern 4th and 4.5[th] generation aircraft, with the remainder being older, less modern 2nd and 3rd generation aircraft.[27]

It is still not clear as whether the PLAAF intends to replace its ageing H-6 (Tu-16 Badger) intermediate-range bomber fleet with the Russian Tu-22M and Tu-95 bombers, which were offered to China in 2005. The Modernisation of its ground-attack forces continued with the delivery of more JH-7 fighter-bombers. So far at least five JH-7 regiments (3 in the Naval Aviation and 2 in the Air Force) have been identified. At the same time, older aircraft are being upgraded with new systems and capabilities to extend their service life. The ageing H-6 bomber is now being used for carrying the land-attack cruise missile, while the obsolete Q-5 attackers are given the ability to deliver laser-guided munitions.

In 1999, PLAAF operated over 3500 combat aircraft comprising mainly the J-6 (MiG-19 equivalent) and the J-7 (based on the MiG-21). A deal with Russia saw the induction of 100 Su-27 fighters. PLAAF also had in its inventory the H-6(Tu-16 based) bombers. China had no precision-guided munitions (PGMs) and only the Su-27 was BVR compatible. Modernisation of the PLAAF has been propelled by China's astounding economic growth. The 21st century has witnessed the acquisition of 105 Su-30MKK from 2000 to 2003 and 100 upgraded Su-30MKK2 in 2004. China produced more than 200 J-11s from 2002 onwards. The PLAAF also bought a total of 126 Su-27SK/UBK in three batches. The production of the J-10 combat aircraft began in 2002 and 1200 are on order. The H-6 bombers (Tu-16 Badger) were converted into flight refueling aircraft. In 2005, the PLAAF unveiled plans to acquire 70 Il-76 transport aircraft and 30 Il-78 tankers to significantly upgrade strategic airlift capability and offer extended range to the fighter force. The US Department of Defence has reported that Su-27 SKs are being upgraded to the multi-role Su-27 SMK status.

The PLAAF is also organizing a combat air wing for a future aircraft carrier group, possibly based on the Su-33, which is a carrier capable variant of the Su-27. Many existing fighters are being upgraded, some for night maritime strike role, permitting carriage of Russian weapons, including Kh-31A anti-radiation cruise missile and KAB-500 laser-guided munitions. PLAAF's aim is to have a primarily fourth generation air force. JH-7/7A will be the backbone of the precision strike force with large numbers of J-10 and J-11 in the air superiority role. The interceptor role will be undertaken by the JF-17 which is under production now in China. The transport force will have Il-76, Il-78 and Y-9 aircraft. China has a variety of helicopters and other aircraft to undertake specialist missions and routine tasks. With a fast developing C4ISR and its shift to joint operations, the Chinese military will be a formidable force to reckon with even by a well prepared adversary. In this process of Modernisation the PLAAF has improved exponentially, though it has yet to be tested in actual operations.

A leading Chinese airpower strategist, Lt Gen Liu Yazhou, has written that the world's major air forces have progressed from ancillary to decisive players. Thus, he says, the People's Liberation Army Air Force (PLAAF) must build its strategic capabilities to enter the top tier of airpower nations[28].

The trends of Modernisation of the PLAAF from 2009 onwards are attached as Appendix 4.

15th AB Corps

The first Chinese Paratroop unit was raised by the KMT in January 1944, which was subsequently expanded to three groups with a Command established under the KMT Air force. After the founding of the PRC, in July 1950, the PLA raised "The Air Force Marines First Brigade" the first air borne unit, under PLAAF. This brigade in September 1951 was renamed the "Air Force First Marine Division" and subsequently renamed "Airborne Division" in 1957. The "Airborne Corps" was established in early 1960s under the PLAAF.[29] From this time onwards the 15th AB Corps has undergone continuous improvement in its organization, structure, weapons and equipment. It has enhanced the levels of military training and air borne combat capability in consonance with the "Core Missions" and to meet the requirements to "fight and win" Local wars under the conditions of informationalization.

From 2000 onwards the transformation of the AB corps from a sole parachuting combat force to a synthesis of ground and air combat forces is very evident. In some very visible steps towards this the mechanization of their primary and combat equipment has been made suitable for air borne operations, battle field mobility has become three dimensional and their combat capability has expanded towards air mobile operations, special air operations and ground assaults. The present organization of the 15th AB Corps is as under;[30]

Table 3.2: Organization of 15 AB Corps

Unit	Numbers	Remarks
Special Forces		
• SF Unit	1	
Manoeuvre		
Reconnaissance		
• Recce Regt	1	
Air Manoeuvre		
• AB Division	2	2 AB Regt, 1 Arty Regt
• AB Division	1	1 AB Regt, 1 Arty Regt
Aviation		
• Helicopter Regt	1	• **CSAR**; *Z-8KA – 08.* • **MRH**; *Z-9WZ – 22.*
Combat Support		
• Signals Group	1	
Combat Service Support		
• Logistics Group	1	

Present Locations of 15th AB Corps and Subordinate Units[31].   The 15th AB Corps has its headquarters at Xiaogan, Hubei province and the three divisions located as under;

- 43rd AB Division – Kaifeng, Henan Province.

- 44th AB Division – Guangshui, Hubei Province.

- 45th AB Division – Wuhan, Hubei Province.

The AB formations rely on transport aircrafts of Guangzhou MRAF's 13th Air Division at Wuhan, Hubei province for airlift support.

Present Deployment of the PLAAF[32]

Map 3.1: Present Deployment of PLAAF

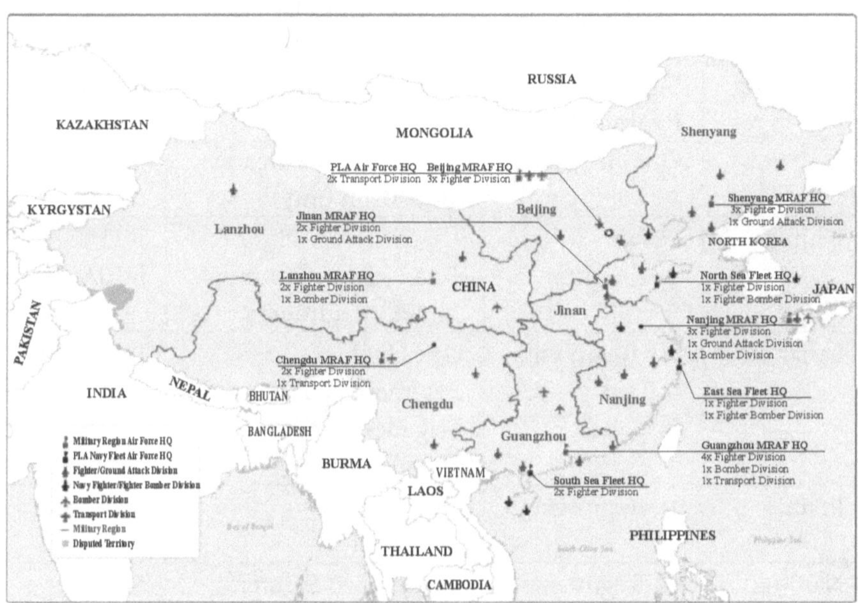

(Source: www.lib.utexas.edu/maps/middle_east_and_asia/china_major_air_and_naval_units-2012.png)

The detailed deployment of the PLAAF as extracted from The Military Balance 2013 is at Appendix 3. A summary of the same is given in the table below:

Table 3.3: Summary of Deployment PLAAF

MRAF	Total No of Regiments & Brigades	Regiments & Brigades with 4th & 4.5 Generation Combat Aircrafts	Type of Aircrafts
Shenyang	Eight Regiments & One Brigade	Two (Fighter & Fighter Ground Attack) Regiments & Brigade (Squadron only)	J-11B & J-10A
Beijing	Seven Regiments	Three (Two Fighter & One Fighter Ground Attack) Regiments	J-11 & J-10A
Lanzhou	Six Regiments & Two Brigades	One (Fighter) Regiment & Two Brigades (One Squadron each)	J-11 & J-11B
Jinan	10 Regiments	One (Fighter) Regiment	Su-27SK
Nanjing	10 Regiments & Two Brigades	Three (Two Fighter Ground Attack & One Fighter) Regiments & Two Brigades (Fighter Ground Attack ; Squadron each)	Su-30 MKK, J-11 & J-10A
Guang-zhou	Nine Regiments & Two Brigades	Four Regiments (Three Fighter Ground Attack & One Fighter)	J-10A, J-11 & Su-30MKK
Chengdu	Four Regiments	Two (One Fighter Ground Attack & One Fighter) Regiments	J-10 & J-11

An Assessment

- In 2013, PLAAF had a total of 82 regiments and brigades out of them only 21 regiments and brigades have 4th or 4.5 generation fighter and fighter ground attack aircrafts.

- Out of 54 regiments only 16 have 4th or 4.5 generation fighter and fighter ground attack aircrafts where as five brigades out of seven have 4th or 4.5 generation fighter and fighter ground attack aircrafts.

- About half the holding of the 4th or 4.5 generation fighter and fighter ground attack aircrafts is with the Guangzhou, Nanjing and Jinan MRAFs.

- Lanzhou and Chengdu have a total 12 regiments and brigades in their Orbat, out of which five regiments and brigades hold 4th or 4.5 generation fighter and fighter ground attack aircrafts.

Analysis of the Modernisation Trends 2009 Onwards. The detailed trends of Modernisation in terms of equipment and formations are attached as Appendix 3. The analyses of these trends indicate the following:

- The MRAFs, today have a balanced mix of the modern and old fighter aircrafts in its inventory. J-10, J-11, Su-27 and Su-30 are the modern aircrafts in its inventory, of which each MRAF divisions hold a regiment.

- The bomber fleet is purely based on the upgrades of the H-6, a variant of the Soviet Tu-16 which was introduced in the PLAAF in 1959.

- Each MRAF, today has one division with modern 4th/ 4.5 generation fighter aircrafts. The Su-30 MKK are on the ORBAT of the Guangzhou and Nanjing MRAF. The other MRAFs have the Su-27 and the J-11s.

- The force multipliers are also deployed with a bias to Taiwan and the Western Pacific.

Recruitment and Training

Beyond these points, the PLAAF has put new emphasis on recruiting and training pilots. The MND announced last fall that the PLAAF had joined with Tsinghua University, a prestigious technical institute, to set up a program intended to draw the best and the brightest into the air service. The first group of pilot cadets was required to meet the academic standards of civilian applicants and the physical standards of the PLAAF. Those accepted will spend three years as undergraduates at Tsinghua, and then transfer to the Aviation University of the Air Force for their final year of college. This effort would be analogous to the USAF joining with MIT or Cal Tech to educate young officers—and potentially, someday, Air Force generals.

Meantime, the Aviation University welcomed new pilot cadets into the "blue sky phalanx." Xu Qiliang, the PLAAF commander, and Gen Deng Changyou, the PLAAF's political commissar, flew from Beijing to Jilin Province, to attend the ceremony in September 2011. Singled out for a welcome was Wang Xu, the first pilot cadet recruited from Tsinghua University to Aviation University. He has been selected to attend Aviation University for further studies.

In flight training, today's PLAAF reportedly gives pilots 200 hours a year in the air, a striking increase from the fewer than 24 hours a year during the depths of the Cultural Revolution. In this respect, the PLAAF is approaching the standard set by USAF. China's days of fielding obsolete air forces with poor training and outdated doctrine have clearly come to an end.

Constrains in the Force Development of the PLAAF

Aero-Engines. The major problem that China faces is that of aero-engines. Presently, almost all her modern aircraft, including J-10, J-11 and JF-17/FC-1, are powered by Russian or Ukrainian designed engines – either bought outright or license-manufactured. China's J-31 stealth fighter reportedly flew with Russian RD-93 engines as did the first flight of J-20. China has been trying to develop high performance aero-engines but has not had the requisite success. Chinese purchase of Su-35 was linked to its purchase of S-117 engines for its J-20 program. A Beijing-based PLA senior Colonel, who requested anonymity, said: "We decided to buy the Su-35 because it's a fact that our home-made engines have failed to measure up to the Russian products." Turbofan engines are technologically advanced and their manufacturers are limited. Due to secrecy that surrounds Chinese defence industry, one is not sure how long would the Chinese continue their dependence on imports. China cannot become a front-line air power unless she masters aero-engine technologies and is able to meet her domestic and export customers' requirements. The problem is not restricted to engines for combat aircraft but also its transport fleet as well; C-919 will have American engines while its Y-20's first flight was with Russian engines. Even if there is a break-through, it would take a few years before production is streamlined as integration of engines with the airframe is a long and time consuming process.

Aircraft Radars. Another constrain is that of active electronically scanned array (AESA) radar which has now become standard upgrade for the western fourth-generation combat aircraft and the standard equipment for fifth-generation fighters. According to Russian sources, the Chinese have not produced an AESA design to date – the Russians showcased their Zhuk-AE AESA originally designed for the MiG-35 at Air Show China in November 2012 to generate Chinese interest in a potential cooperative effort to develop an AESA radar set for one or more advanced fighters under development in China. Radar is the heart of any fighter and again Chinese industry has a lot of catching up to do to come up to the industry's standard in this field. There are some unconfirmed reports that J-15 has a Chinese designed AESA radar; if true, this speaks highly of the Chinese aircraft industry. In addition, there have been unconfirmed reports that Russia had supplied AESA radar for Russian-supplied Su-30MKK aircraft. If the reports are correct, it should assist China in finding a solution to her problems in her quest for the vital indigenous AESA. One would have to wait for confirmation of the success and thereafter assess the timeframe for retro-modification of the existing fleet.

Electronic Warfare. Another grey area where no formal assessments are available is the technological sophistication of the Chinese airborne electronic warfare suites. It has been generally accepted that the Americans and the European are way ahead of the Russians and since the Chinese have either Russian equipment or indigenous EW equipment based on Russian designs; one cannot make an accurate assessment of their efficacy in actual war scenario. When asked about the effectiveness of their EW suites, one of the USAF Chiefs replied that one can find an answer to this question only after a war and there is no way a proper assessment can be made in peace time! It is clear that the Chinese aircraft industry has a long path ahead if it wants to match the best in the world – the advances in the American and the Russian industry continue meanwhile; one expects that the technological gap would reduce after a decade or so. There may be no paucity of funds but research and development of sophisticated technology is time consuming and serial production thereafter takes time.

PLAAF in 2025

Likely Strength in 2025. A few points that need to be borne in mind when assessing the force level of the PLAAF in the next two decades or so. It can be said with near certainty that China would not increase her strength to 4,000 combat aircraft as was the case in 1960s and 1970s as this figure is just not affordable. Currently, the average flying for pilots varies between 100 to 150 hours/year instead of the ideal and desirable 200 hours/year. It was assessed in the late 1990s and early years of decade of 2,000 that China would downsize to around 1,000 fourth/fifth-generation aircraft. The present inventory level is around 1,800 aircraft, including third-generation aircraft. There are additional 300 combat aircraft with the PL Navy. There would normally be a case for reduction in the total size of inventory, though it is unlikely that outstanding issues that China faces (like Japan, Taiwan, Philippines, South Korea, India and most importantly the United States) would get resolved in the next 10-15 years. It is assessed that the PLAAF would maintain its present inventory level though its ageing aircraft would be replaced with modern and state-of-the-art aircraft – a combination of fourth- and fifth-generation aircraft. If there is any increase, the total is unlikely to go beyond 2000 combat aircraft.

Trends up to 2025

- The development and employment of 'Stealth and Counter Stealth' will be the focus of the PLAAF. The first Chengdu J-20 stealth fighters will be operationalized by the PLAAF by around 2020. Around the same time the introduction of Shenyang J-31 or J-21 may also take place in the PLAAF only or in PLAN also. By 2025 the PLAAF will have in its inventory a host of 4th & 4.5 generation aircrafts like the J-20, J-31, Su-35, Su-30 and Su-27 and these will be the mainstay of the PLAAF with an effective counter stealth capability.

- A number of force multipliers will form part of the PLAAF with Y-20 forming the basic platform for these force multipliers. By 2025, China will be able to able to produce approximately 200 to 250 Y-20 transport aircrafts, which will be able to cater for the requirements of the PLAAF as far as force multipliers and force projection capabilities are concerned.

- By 2025 China will have a full fledged Integrated Air & Space

Defence architecture in place.

- A number of UAVs & UCAVs will be in service will the PLAAF catering for all tasking and altitudes. Out of the total inventory, there will be a large percentage of stealth UAVs and UCAVs. Also by 2025 China may develop capability to land UAVs and UCAVs on her aircraft carriers and launch them from rockets and missiles.

Major changes are likely in the PLAAF in coming years as the thinking in the Chinese political and military leadership changes. If some or all of these changes occur, the already rising influence of the PLAAF will most likely climb higher. This will affect priorities, budgets, procurement of weapons, and assignment of senior leaders—but for this an important precursor will be the ability of the PLAAF to break the traditional dominance of China's military forces by the Army.

Endnotes

1   The Military Balance 2014, IISS, Routledge, London, 2014, p. 235.

2   Lu Xiaoping, "The PLA Air Force", China Intercontinental Press, Beijing, 2011, p. 3.

3   Ibid, p. 13.

4   Evolution of PLAAF doctrine/ training, Information Dissemination, August 31, 2011, accessed from www.informationdissemination.net/2011/08/evolution-of-plaaf-doctrinetraining.html .

5   Jiang Guocheng, "Building an Offensive and Decisive PLAAF: A Critical Review of Lt Gen Liu Yazhou's *The Centenary of the Air Force*", Air & Space Power Journal, Summer 2010, June 1, 2010 accessed from http://www.airpower.maxwell.af.mil/airchronicles/apj/apj10/sum10/11jiang.html on July 07, 2014.

6   Ibid.

7   Richard Halloran, A Revolution for China's Air force , Air force Magazine, February 2012

8   Information Office of the State Council, *"The Diversified Employment of China's Armed Forces"* April 2013, Beijing.

9   Ibid.

10 Office of the Secretary of Defence, Annual Report to the Congress: Military

and Security Developments Involving the People's Republic of China 2013

11 Cliff Roger, Fei John, Hagen Jeff, Hague Elizabeth, Heginbotham Eric, Stillion, "Shaking the Heavens and Splitting the Earth", Rand Corporation, Pittsburg, USA, 2011, p. 7.

12 National Air & Space Intelligence Center, "People's Liberation Air force", NASIC Public Affairs Office, Ohio, August 2010, p. 1.

13 Phil Stewart, Gates: China confirms stealth jet test-flight, Reuters, January 11, 2011, accessed from http://www.reuters.com/article/2011/01/11/us-china-defence-fighter-idUSTRE70A19B20110111 on May 13, 2014.

14 Ulman Wayne A., "China's Emergent Military Aerospace and Commercial Aviation Capabilities", Hearing before the US-China Economic and Security Review Commission, May 20, 2010, Washington DC, p. 57.

15 Dong Wenxian, "Part I of the 'Strategic Air Force' Series: The Expansion of National Strategic Space Calls for a Strategic Air Force", Air Force News February 2, 2008, p. 2.

16 Ibid.

17 Pollpeter Kevin, no. 5.

18 The Military Balance 2014, IISS, Routledge, London, 2014, p. 236.

19 Pollpeter Kevin, "PLAAF and Integration of Air and Space Power" in Richard P. Hallion, Roger Cliff and Phillip C. Saunders (ed.) The Chinese Air force: Evolving Concepts, Roles and Capabilities, National Defence University Press, Washington DC, 2012, pp. 172-173.

20 Il 76 – Production, www.globalsecurity.com, accessed from http://www.globalsecurity.org/military/world/russia/il-76-production.htm on July 12, 2014.

21 Test flight frequency of Y-20 heavy-duty transport aircraft sets new record, China Military Online, March 5, 2014, accessed from http://eng.chinamil.com.cn/news-channels/china-military-news/2014-03/05/content_5796123.htm on May 13, 2014.

22 Lu Xiaoping, "The PLA Air Force", China Intercontinental Press, Beijing, 2011, p. 105.

23 The Military Balance 2014, IISS, Routledge, London, 2014, p. 236.

24 Ibid.

25 Cliff Roger, Fei John, Hagen Jeff, Hague Elizabeth, Heginbotham Eric, Stillion, "Shaking the Heavens and Splitting the Earth", Rand Corporation, Pittsburg, USA, 2011p. 125.

26 Ibid.

27 Ryan, Colonel Mick, India – China in 2030: A Net Assessment of the Competition between Two Rising Powers

28 Jiang Guocheng, "Building an Offensive and Decisive PLAAF: A Critical Review of Lt Gen Liu Yazhou's *The Centenary of the Air Force*", Air & Space Power Journal, Summer 2010, June 1, 2010 accessed from http://www.airpower.maxwell.af.mil/airchronicles/apj/apj10/sum10/11jiang.html on July 07, 2014.

29 Lu Xiaoping, "The PLA Air Force", China Intercontinental Press, Beijing, 2011, p. 104.

30 The Military Balance 2014, IISS, Routledge, London, 2014, p. 236.

31 China's Air Force 2010, National Air and Space Intelligence Center, Ohio, US, August 1, 2010, p. 96.

32 The data has been extracted from *The Military Balance 2014* published by IISS in 2014.

4

PLA Navy (PLAN) Modernisation

"Without a decisive naval force we can do nothing definitive, and with it, everything honorable and glorious."

- President George Washington

Starting the 1990s the PLAN has transformed itself from a large fleet of low capability – single mission platforms to a leaner navy with modern and multi mission platforms. This journey of PLAN in fact started many years ago with its founding on April 23rd, 1949 with the establishment of the East China Military in Baima Temple in Taizhou, Jiangsu Province. In February 1953, Chairman Mao embarked on warship *Changjiang* of PLAN to inspect. He spent four days aboard and while disembarking wrote "We must establish a powerful naval force to fight against the imperialist aggressions"[1]. The PLAN had its first operational experience in the capture of the Yijiangshan islands from January 18 to January 20, 1955 during the First Taiwan Strait Crisis. In this operation a total of five landing transport fleets composed of 46 frigates, gunboats and escort boats took control of the Yijiangshan islands from the Nationalists, in coordination with the PLAA and PLAAF[2]. In the summer of 1979 Deng Xiaoping, while aboard the first indigenous missile destroyer, *Jinan*, said "the ocean is not a narrow city moat and the navy should not be the defenders of a city. To construct and establish a strong and wealthy China, we should head for the world and face the ocean"[3]. From here on the course of the vessels of the PLAN has expanded from the Chinese coastlines to the oceans. Today the PLAN has the largest force of major combatants, submarines and amphibious warfare ships in Asia. PLAN's naval forces include 77 principal surface combatants, including an aircraft carrier, more than 60 submarines, 55 medium and large amphibious ships and approximately 85 missile equipped small combatants[4].

The White Paper on 'The Diversified Employment of China's Armed Forces, 2013' enunciates and elucidates the role, composition and the future trends of development of PLA Navy (PLAN) as;

'The PLA Navy (PLAN) is China's mainstay for operations at sea, and is responsible for safeguarding its maritime security and maintaining its sovereignty over its territorial seas along with its maritime rights and interests. The PLAN is composed of the submarine, surface vessel, naval aviation, marine corps and coastal defence arms. In line with the requirements of its offshore defence strategy, the PLAN endeavors to accelerate the Modernisation of its forces for comprehensive offshore operations, develop advanced submarines, destroyers and frigates, and improve integrated electronic and information systems. Furthermore, it develops blue-water capabilities of conducting mobile operations, carrying out international cooperation, and countering non-traditional security threats, and enhances its capabilities of strategic deterrence and counterattack. Currently, the PLAN has a total strength of 235,000 officers and men, and commands three fleets, namely, the Beihai Fleet, the Donghai Fleet and the Nanhai Fleet. Each fleet has fleet aviation headquarters, support bases, flotillas and maritime garrison commands, as well as aviation divisions and marine brigades. In September 2012, China's first aircraft carrier Liaoning was commissioned into the PLAN. China's development of an aircraft carrier has a profound impact on building a strong PLAN and safeguarding maritime security.'

Transformation of PLAN

By the end of 20th century, the PLAN began moving towards an off-shore defensive strategy, relying on more out-of-area operations, away from the traditional focus on territorial waters. Offshore defence, alternatively translated as "Near Sea Defence", is a 'regional' strategy that does not advocate replicating the US or the Soviet style 'Blue Water' Naval capabilities, but calls for naval capabilities suited for China's specific regional maritime interests. The direction for research towards Offshore Defence was given by Admiral Liu Huaqing in 1982[5]. The area of China's immediate interest was defined by Admiral Liu. It includes the Yellow, East China, South China Seas, the seas around the Spratly Islands and Taiwan, areas inside and outside the Okinawa Island chain, as well as the northern part of the Pacific Ocean. Chinese scholars and naval strategists now advocate a new strategy for the 21st century termed 'distant sea defence' which does not bind operations geographically but would be defined according to China's maritime needs.'[6]

This shift towards blue waters beyond the 'first island chain' was accompanied by a large number of overseas ship visits. Between 1985 and 2006, PLAN vessels visited 18 Asian-Pacific, four South American, eight European, three African and three North American nations. In 2003, the PLAN conducted its first joint naval exercises during separate visits to Pakistan and India. Bi-lateral naval exercises were also conducted with the French, British, Australian, Canadian, Philippine, and United States navies.

Modernisation Trends of the PLAN

Enhancement of C4ISR and in terms of logistics capabilities are at the heart of China's Modernisation strategy. The PLA has increased efforts to create an advanced C4ISR network among PLA branches and services by shifting to IT-enabled weapon systems, proliferation of information warfare units and efforts to recruit highly qualified civilian IT experts. PLA leaders understand that conducting 'integrated joint operations' is virtually impossible without effective, decentralized C4ISR networks, and they have identified the PLA's deficiencies in this sector as a key stumbling block to efforts at joint operations.[7]

- The PLA Navy's Modernisation and development, based on changes in national security doctrine, is reflected in a sustained shift from a navy of large numbers of single-mission vessels, mostly patrol craft. The navy now has significant numbers of modern major combatants, as well as modernized patrol craft.

- The PLAN has also modernized its weapon systems in a manner similar to the PLAA. Large numbers of obsolete vessels, mostly coastal combatants, have been discarded and replaced by modernized, imported and indigenously-manufactured ocean-going combatants.

- The PLAN is replacing its patrol craft with modern variants such as the *Houbei,* which has a wave-piercing hull design and can carry eight anti-ship cruise missiles.

- New diesel and nuclear-powered submarines have given a significant boost to the PLA Navy's undersea capabilities, and have quadrupled the number of submarines capable of firing Anti-Ship Cruise Missiles (ASCM).

- The latest combatants feature mid and long range surface to air missiles and the Luyang II DDG possesses a sophisticated phased array radar system similar to western AEGIS radars.[8]

- Underway replenishment capabilities have increased by 67 percent.

- Sea trials of the aircraft carrier *Liaoning* (ex-*Varyag*), as well as flight-testing of the J-15 (Su-33) carrier-fighter, indicate a shift toward enhanced power-projection capabilities. This development also underscores the remarkable similarity between the development of the former Soviet Navy and the PLA Navy. If the analogy is sustained, it would indicate a medium-term role for the Chinese aircraft carriers in support of anti-access operations within the First and Second Island Chains rather than the imminent formation of Carrier Battle Groups patterned on the US Navy. The Chinese aircraft carriers would also make a substantial contribution to the PLA Navy's peacetime maritime diplomacy capabilities.

A detailed analysis of these trends is as under:-

Command, Control, Communications, Computers, Information, Intelligence, Surveillance and Reconnaissance (C⁴I²SR) Systems.

China is deploying a new generation of C^4I^2SR systems and networks, including communications networks, data links, intelligence collection systems, navigation satellites, and information fusion systems. These C^4I^2SR capabilities will be the key factor in the PLA's ability to coordinate operations within and among its ground, air, naval, and missile forces; detect and track foreign military activities throughout the Western Pacific and Indian Ocean Region (IOR); and conduct time-sensitive, long-range strikes with its ballistic and cruise missiles.[9]

<u>Anti-ship Ballistic Missiles (ASBM)</u>. In 2010, China deployed the DF-21D ASBM. The DF-21D, which has a range exceeding 810 nautical miles (nm), provides Beijing with the ability to threaten large surface ships, such as aircraft carriers, throughout the Western Pacific and IOR. China is fielding additional DF-21Ds and may be developing a variant capable of reaching ships at longer ranges.[10]

Cruise Missiles. China has fielded multiple classes of anti-ship cruise missiles (ASCMs) aboard the PLA Navy's growing inventory of mobile launch platforms (submarines, surface combatants, and aircraft) and to land-based units along the Chinese coast. These include Russian ASCMs, such as the SS-N-22 (130 nm) and the SS-N-27 (120 nm), as well as Chinese ASCMs, such as the YJ-62 (150 nm), the YJ-83 (95 nm), the YJ-82 (65 nm), and the YJ-8A (65 nm). China also has deployed multiple classes of surface-to-air missiles (SAMs) on its modern destroyers and frigates. These include Russian SAMs, such as the SA-N-20 (80 nm) and the SA-N-7 (12-20 nm), as well Chinese SAMs, such as the HHQ-9 (55 nm) and the HHQ-16 (20-40 nm).

China's ASCM and SAM research, development, and production has accelerated since the early 2000s, and future missiles will have longer ranges and enhanced survivability, lethality, and employment flexibility. China also is developing an indigenous supersonic ASCM.[11]

Although the PLA Navy has a significant inventory of ASCMs, it currently does not have the ability to strike land targets. However, the PLA Navy recently began to develop submarine- and surface-launched land attack cruise missiles (LACMs). A future sea-based LACM will complement the PLA's air- and ground-based LACMs and ballistic missiles, enhancing Beijing's flexibility for attacking land targets throughout the Western Pacific and beyond.[12]

Nuclear Ballistic Missile Submarines. China's JL-2 submarine-launched ballistic missile (SLBM) likely will reach initial operational capability by late 2013[13]. The JL-2, when mated with the PLA Navy's JIN nuclear ballistic missile submarine (SSBN), will give China its first credible sea-based nuclear deterrent. The JIN SSBN/JL-2 weapon system will have a range of approximately 4,000 nm, allowing the PLA Navy to target entire Asia, Europe and also targets up to continental United States from China's littoral waters.[14] China has deployed three JIN SSBNs and plans to field as many as five by 2020.[15] China also is developing its next generation SSBN, the Type 096,[16] which probably will improve the range, mobility, stealth, and lethality of the PLA Navy's nuclear deterrent.

Aircraft Carriers. In 2012, China commissioned its first aircraft carrier, the *Liaoning* aircraft carrier (CV), after approximately six years of renovation work on the Ukrainian-built hull and one year of sea trials. The prestige and political value of China's aircraft carrier program currently outweigh its military utility.[17] Officially classified as a research and training vessel, the *Liaoning* CV will serve largely to build the PLA Navy's proficiency in aircraft carrier operations until the navy develops a carrier aviation capability in 2015 or later.[18] A fully operational *Liaoning* CV with an attached air wing will allow Beijing to project more air power against less advanced navies in the region and improve the PLA Navy's ability to conduct a range of missions, including airborne early warning, command and control, and humanitarian assistance. China likely intends to follow the *Liaoning* CV with at least two domestically produced hulls,[19] the first of which appears to be under construction at Shanghai's Changxing Island Shipyard.[20] This carrier could become operational before 2020.[21]

Submarine and Surface Forces. In 1990, the PLA Navy had large numbers of antiquated and low-capability submarines and coastal patrol boats. The PLA Navy since then has shed many of its legacy units and acquired – through a combination of indigenous development and Russian acquisition – more stealthy and lethal submarines and larger and more capable surface combatants. Thus, while the PLA Navy's total number of submarines and surface ships declined from 1990 to 2010, the force experienced significant qualitative improvements over this timeframe. China continues to steadily increase its inventory of modern submarines and surface combatants, as well as large amphibious ships and logistics ships. China is building at least seven major classes of submarines and surface ships simultaneously, more than any other country in the world except for the United States.[22]

- In 2012, China began building four "improved variants" of its *Shang* nuclear attack submarine (SSN). China also continues production of the *Yuan* conventional submarine, some of which will include an air-independent power (AIP) system that allows for extended duration operations, and the *Jin* SSBN (discussed above). Furthermore, China is developing two new classes of nuclear submarines; the Type 095 guided-missile attack submarine (SSGN) and the Type 096 SSBN – and may jointly develop four advanced conventional submarines with Russia.[23] China's growing inventory of nuclear and conventional submarines will significantly enhance China's ability to strike opposing surface ships throughout the Western Pacific and IOR to protect future nuclear deterrent patrols and aircraft carrier task groups.

- In 2012, China launched two new classes of naval combatants –

the *Luyang III* guided-missile destroyer (DDG) and the *Jiangdao* corvette – and resumed construction of the *Luyang II* DDG after a brief hiatus. China also continues serial production of the *Jiangkai II* guided-missile frigate and it is estimated that most of these units will be available to PLAN by 2015. The PLA Navy's expanding and modernizing surface force will improve Beijing's ability to project power in the Taiwan Strait, East China Sea, South China Sea, Western Pacific, and IOR and fulfill the PLA Navy's growing missions in China's distant seas, such as defence of SLOCs, humanitarian assistance, and counter-piracy.

- In 2012, the PLA Navy commissioned two new *Yuzhao* amphibious transport docks (LPD), bringing its total number of LPDs to three. The *Yuzhao* LPD can carry a mix of air-cushion landing craft, amphibious armoured vehicles, helicopters, and marines. This provides the PLA Navy with enhanced flexibility for a range of missions, such as amphibious assault, humanitarian assistance, and counter-piracy. China may build additional *Yuzhao* LPDs and plans to introduce a new landing helicopter assault ship, called the Type 081, by 2018[24].

- In 2013, China added two upgraded *Fuchi* auxiliary replenishment oilers (AORs) to its fleet, bringing its total number of AORs from five to seven. The increased number of naval support ships better equips the PLA Navy's modern surface ships and future aircraft carriers with the means to sustain high-tempo operations at longer ranges.[25]

According to China military analysts Andrew Erickson and Gabe Collins, "...by 2015, China will likely be second globally in numbers of large warships built and commissioned since the Cold War's end... by 2020, barring a U.S. naval renaissance, it is possible that China will become the world's leading military shipbuilder in terms of numbers of submarines, surface combatants and other naval surface vessels produced per year... [and] with respect to overall shipbuilding, China will likely reach 2013 Russian technical-proficiency levels by 2020 and may even reach 2013 U.S. technical levels by 2030."[26]

As part of the Modernisation process, the PLA Navy has revised its manpower policies. In addition to reducing manpower, the PLA Navy now conducts exercises and deployments to develop skills necessary for offshore defence and for gaining experience that is at a premium in a service with little combat experience. These measures are reflected in increased ability to undertake anti-surface warfare (ASuW), naval air defence and force projection missions. However, the PLA Navy continues to lag behind in its anti-submarine warfare capability. The PLAN seems to be conscious of this weakness and has increased the numbers of ASW helicopters to mitigate this deficiency. Overall, the PLA Navy's Modernisation has clearly augmented China's naval capabilities and improved the Navy's ability to handle regional contingencies envisaged in the Local Wars doctrine.

PLA Navy's effort to develop a professional force rests on three pillars: professional NCOs, academically qualified officers, and additional promotion and educational opportunities for enlisted personnel. Since 1999, the PLA Navy has reduced the conscription obligation from four to two years, while doubling maximum service for NCOs from 15 to 30. NCOs are also taking over many jobs previously performed by officers. Meanwhile, the numbers of officers are shrinking as the PLA Navy seeks fewer but better qualified personnel. Towards this end, the PLAN has increased academic scholarships for reserve officers, increased technical training in the fleet and stated looking for officers with advanced degrees. To upgrade skills among serving personnel, the PLAN has also introduced on-the-job, short-term, and on-line training programs.

The table below highlights changes in quantitative terms. The period 1990-2012 has seen the PLAN rapidly discarding obsolete coastal naval assets and procuring advanced major surface combatants through imports if necessary. The PLA Navy has thus been transformed into a force capable of creating a hazardous environment, in East Asia, for any possible adversary including the United States.

Transition from 2007 onwards[27]. The transition of the PLAN from 2007 to 2013 is attached as Appendix 5.

The Present Deployment of PLAN

Since the early 1990s, PLAN has undergone rapid Modernisation and is now the second largest navy in the world, second only to the United States Navy. Today, PLAN has a personnel strength of over 235,000 that also includes:

- 35,000-strong Coastal Defence Force

- 56,000-strong PLA Marine Corps

- 56,000-strong PLA Naval Air Force, operating several hundred land-based aircraft and ship-based helicopters.

The PLA Navy is organized into several departments for purposes of command, control and coordination and the main operating forces are organized into three fleets whose geographic locations along with their principal combatants is given on the map below.[28]

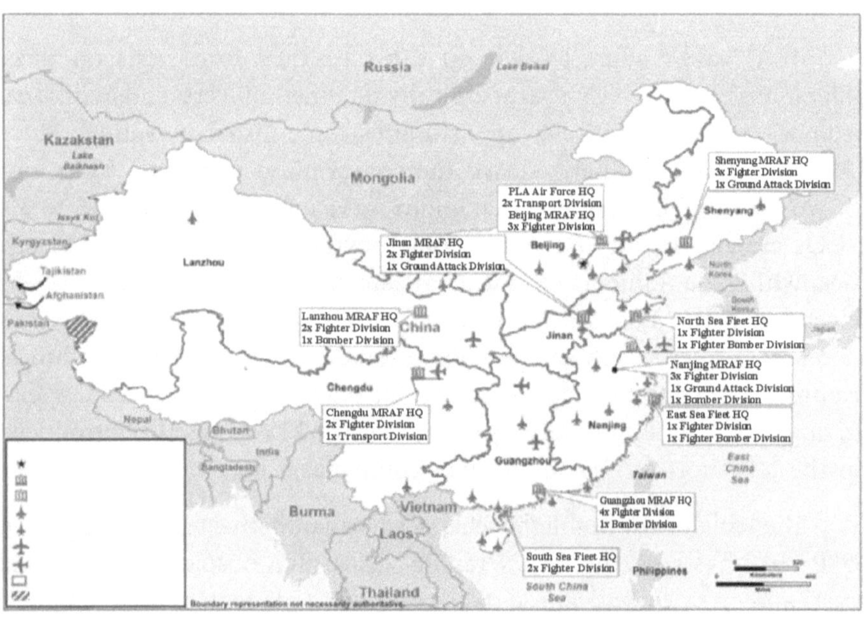

Map 4.1: Present Deployment of PLAN

(Source: Annual Report to Congress- Military and Security Developments Involving the People's Republic of China 2013)

North Sea Fleet

This fleet is responsible for maritime defence from the border of Democratic People's Republic of Korea (Yalu River) in North to Lianyungang in South. This equates to the Shenyang, Beijing and Jinan MACs and to seaward. The HQ of the fleet is at Qingdao with support bases at Lushun and Qingdao. In addition there are nine coastal-defence districts in its area of responsibility. The details of the fleet are as under[29];

Table 4.1: North Sea Fleet Details

Submarines	
Nuclear Powered Ballistic Missile Submarine (SSBN)	3
Nuclear Powered Attack Submarine (SSN)	3
Attack Submarine with Anti-Submarine Warfare capability (hunter killer) (SSK)	20
Major Surface Combatants	
Destroyer with Anti-Ship Missile, hanger & SAM (DDGHM)	2
Destroyer with Anti-Ship Missile & SAM (DGM)	2
Frigate with Anti-Ship Missile, hanger & SAM (FFGHM)	5
Frigate with Anti-Ship Missile & SAM FFGM	2
Frigate with Anti-Ship Missile & hanger (FFGH)	1
Frigate with Anti-Ship Missile (FFG)	6
Mine Layer (ML)	1
Patrol Craft Fast with Guided Missile(PCFG)/ Patrol Craft with Guided Missile (PCG)	22
Patrol Craft Coastal with Anti-Ship Missile (PCC)	28
Landing Ship (LS)	9
Mine Countermeasures Vessel (MCMV)	7

North Sea Fleet Aviation. The details of the Naval Aviation are as under:-

- 5th Naval Air Division

  - ➤ 1 FGA Regiment with JH-7A

  - ➤ 1 Fighter Regiment with J-8F

- Other Forces

  - ➤ 1 EW/ ISR/ AEW Regiment with Y-8J/ JB/ W

- ➢ 1 Maritime Patrol (MP) Regiment withSH-5

- ➢ 1 Helicopter Regiment with AS365; Ka-28; SA 321; Z-9

East Sea Fleet

The East Sea Fleet is responsible for coastal defence form South of Lianyungang to Dongshan. This equates to the Nanjing Military Region, and to seaward. The HQs of the fleet are at Ningbo with support bases at Fujian, Zhoushan and Ningbo. There are seven coastal defence districts in the area of responsibility of the fleet. The details of the fleet are as under[30];

Table 4.2: East Sea Fleet Details

Submarines	
Attack Submarine with Anti-Submarine Warfare capability (hunter killer) (SSK)	17
Major Surface Combatants	
Destroyer with Anti-Ship Missile, hanger & SAM (DDGHM)	6
Frigate with Anti-Ship Missile, hanger & SAM (FFGHM)	14
Frigate with Anti-Ship Missile (FFG)	11
Patrol Craft Fast with Guided Missile(PCFG)/ Patrol Craft with Guided Missile (PCG)	34
Patrol Craft Coastal with Anti-Ship Missile (PCC)	22
Landing Ship (LS)	27
Mine Countermeasures Vessel (MCMV)	22

East Sea Fleet Aviation. The details of the Naval Aviation are as under:-

- • 4th Naval Air Division

  - ➢ 1 FGA Regiment with Su-30 Mk2

  - ➢ 1 FGA Regiment with J-10A

- • 6th Naval Air Division

  - ➢ 2 FGA Regiment with JH-7

  - ➢ 1 Bomber Regiment with H-6G

- • Other Forces: 1 Helicopter Regiment with Mi-8; Ka-28; Ka-31

South Sea Fleet

In addition to the maritime defence from Dongshan to the Vietnam border, this fleet will also be active in the Indian Ocean, a concern for Indian Security establishment. The area of responsibility of the South Sea Fleet corresponds to Guangzhou MR, and to seaward (including Paracel and Spratly Islands). The HQs of the fleet are located at Zuanjiang, with support bases at Yulin and Guangzhou[31].

Table 4.3: South Sea Fleet Details

Submarines	
Nuclear Powered Ballistic Missile Submarine (SSBN)	1
Nuclear Powered Attack Submarine (SSN)	2
Attack Submarine with Anti-Submarine Warfare capability (hunter killer) (SSK)	18
<u>Major Surface Combatants</u>	
Destroyer with Anti-Ship Missile, hanger & SAM (DDGHM)	5
Frigate with Anti-Ship Missile, hanger & SAM (FFGHM)	9
Frigate with Anti-Ship Missile (FFG)	12
Patrol Craft Fast with Guided Missile(PCFG)/ Patrol Craft with Guided Missile (PCG)	42
Patrol Craft Coastal with Anti-Ship Missile (PCC)	20
Landing Platform Dock (LPD)	2
Landing Ship (LS)	51
Mine Countermeasures Vessel (MCMV)	10

South Sea Fleet Aviation. The details of the Naval Aviation are as under:-

- 8th Naval Air Division

  - 1 FGA Regiment with J-11B

  - 1 Bomber Regiment with H-6G

  - 1 Fighter Regiment with J-7E

- 9th Naval Air Division

  - ➢ 12 FGA Regiment with JH-7A

  - ➢ 1 Fighter Regiment with J-8H

- Other Forces: 1 Transport Regiment with Y-7; Y-8; Z-8; Z-8JH/S; Z-9

Certain Trends for the Future

Table 4.4: Future Trends PLAN

PLAN Submarine Orders-of-Battle 1990-2020, Total Numbers

Type	1990	1995	2000	2005	2010	2015	2020
Diesel Attack	88	43	60	51	54	57-62	59-64
Nuclear Attack	4	5	5	6	6	6-8	6-9
Nuclear Ballistic	1	1	1	2	3	3-5	4-5
Total	93	49	66	59	63	66-75	69-78

PLAN Submarine Orders-of-Battle 1990-2020, Approximate Percent Modern

Type	1990	1995	2000	2005	2010	2015	2020
Diesel Attack	0 %	0 %	7 %	40 %	50 %	70 %	75 %
Nuclear Attack	0 %	0 %	0 %	33 %	33 %	70 %	100 %

PLAN Surface Orders-of-Battle 1990-2020, Total Numbers

Type	1990	1995	2000	2005	2010	2015	2020
Aircraft Carriers	0	0	0	0	0	1	1-2
Destroyers	19	18	21	21	25	28-32	30-34
Frigates	37	37	37	43	49	52-56	54-58
Corvettes	0	0	0	0	0	20-25	24-30
Amphibious Ships	58	50	60	43	55	53-55	50-55
Coastal Patrol (Missile)	215	217	100	51	85	85	85
Total	**329**	**322**	**218**	**158**	**214**	**239-254**	**244-264**

PLAN Surface Orders-of-Battle 1990-2020, Approximate Percent Modern

Type	1990	1995	2000	2005	2010	2015	2020
Destroyers	0 %	5 %	20 %	40 %	50 %	70 %	85 %
Frigates	0 %	8 %	25 %	35 %	45 %	70 %	85 %

(Source: U.S.-China Economic and Security Review Commission Staff Research Backgrounder, China's Naval Modernisation and Implications for the United States, August 26, 2013)

PLAN in 2025

The analysis of the factors discussed in the preceding paragraphs leads to a likely force structure of the PLAN by 2025, the details of it as given under;

- The PLAN will be operating minimum of two and a maximum of three conventional (non nuclear powered) aircraft carriers. This will be a major impediment in its global projection ambitions. Regionally PLAN will be capable of achieving the required domination and force projection capabilities in Western Pacific.

- The strategic nuclear submarine force will consist of Jin class (Type 94) and maybe Tang class (Type 96) SSBNs. The Jin class (Type 94) will be the mainstay of her SSBN fleet with about seven to nine in service. By 2025 Tang class (Type 96) SSBN may also see entry into service and PLAN may operate one to two of these SSBNs. Both classes of SSBNs are likely to carry the Julang-2 (Jl-2) or CSS-NX-4 SLBMS with Type 94 carrying 12 SLBMs and Type 96 carrying 24 SLBMs, only if by then Jl-3 SLBM is not operational.

- The conventional submarine force will be based on the Shang Class (Type 93) SSN of which about four to five are likely to be in service. Yuan class submarines with AIP will be the mainstay of the non-nuclear conventional submarine force and about 20-22 submarines will be in service. The remaining submarine force will consist of the aging Kilo and Song class submarines.

- The surface combatant fleet will have the Luyang III (Type 52D) class of destroyers and the Jiangkai II class of frigates as the mainstay. About eight to ten Luyang III destroyers and a similar number of Jiangkai II class of frigates will be in service. The Jiangdao class of corvettes will be primarily for littoral warfare.

- For amphibious operations the Yuzhao Class LPD & Type 81 Landing Helicopter Assault Ship will be the mainstay and these will be supplemented by a no of vessels from the existing holding.

- In the maritime domain, the PLANs Counter Intervention Strategy in 2025 will be based on the DF – 21D Anti Ship Ballistic Missile (ASBM) and a number of Russian and indigenous Cruise Missiles integrated with the Air and Space Defence architecture, other missile forces of the PLASAF and the air defence weapon systems of the PLAAF.

Is the threat of the PLAN with its existing and future capabilities so great, that it would be difficult to handle or counter? According to a senior political scientist at the RAND Corporation, Roger Cliff, "The thing to keep in mind is that, in order for China to successfully attack a U.S. navy ship with a ballistic missile, it must first detect the ship, identify it as a U.S. warship of a type that it wishes to attack (e.g., an aircraft carrier), acquire a precise enough measurement of its location that a missile can be launched at it (i.e., a one-hour old satellite photograph is probably useless, as the ship could be 25 miles away from where it was when the picture was taken), and then provide mid-course updates to the missile. Finally, the warhead must lock onto and home in on the ship. This complicated "kill chain" provides a number of opportunities to defeat the attack. For example, over-the-horizon radars used to detect ships can be jammed, spoofed, or destroyed; smoke and other obscurants can be deployed when an imagery satellite, which follows a predictable orbit, is passing over a formation of ships; the mid-course updates can be jammed; and when the missile locks on to the target its seeker can be jammed or spoofed. Actually intercepting the missile is probably the most difficult thing to do. [...]The missile by itself would be pretty useless. As implied by my response to the previous question, an entire "system of systems" is needed to make it work. Some countries might buy them just to impress their neighbors, but their combat effectiveness would be negligible unless the country also invested in the needed detection, data processing, and communications systems[32]."

A narrative very commonly heard in various discussions and written in numerous publications not necessarily projects the exact trajectory of the PLAN Modernisation. China's military Modernisation program has brought a range of new capabilities to the PLAN. One must realize that new capabilities, such as long range cruise missiles, advanced surveillance systems and data link communication capabilities will serve as significant force multipliers over the next 10 to 15 years. In future even if there is a decrease in the size of PLAN, the overall naval capabilities can be expected increase as forces gain multi-mission capabilities. Proficiency in joint operations, such as integrated air defence will further augment total force capabilities by providing flexible and redundant options to defend naval missions. The pace of military Modernisation has been enabled by China's economic growth, access to advanced technology in the international market and clearly defined national objectives for naval developments. While the pace of Modernisation may slow, China remains committed to continued development of naval capabilities to support China's growing maritime interests.

Endnotes

1   Gao Xiaoxing, The PLA Navy, China Intercontinental Press, Beijing, 2011, p. 1.

2   Ibid, p. 2.

3   Ibid.

4   U.S. Department of Defence, Annual Report to Congress on China's Military Power, Military and Security Developments Involving the People's Republic of China 2014, Washington, DC, 2013), p. 7.

5   The People's Liberation Army Navy: A Modern Navy with Chinese Characteristics, Office of Naval Intelligence, Department of Defence, US Government, August 2009, accessed from http://www.fas.org/agency/oni/pla_navy.pdf on May 13, 2014, p.5.

6   Ibid, p. 6.

7   Ibid, pp. 57-59.

8   Ibid.

9   Shane Bilsborough, "China's Emerging C4ISR Revolution," *Diplomat*, August 13, 2013. *http://thediplomat.com/2013/08/13/chinas-emerging-c4isr-revolution/?all=true*; Kimberly Hsu, *China's Military Unmanned Aerial Vehicle Industry* (Washington, DC: U.S.-China Economic and Security Review Commission, June 13, 2013); Mark Stokes, Jenny Lin, and L.C. Russell, *The Chinese People's Liberation Army Signals Intelligence and*

Cyber Reconnaissance Infrastructure (Arlington, VA: The Project 2049 Institute, November 11, 2012); and Mark Stokes, *China's Evolving Space Capabilities: Implications for U.S. Interests* (Arlington, VA: The Project 2049 Institute, April 26, 2012).

10   Andrew Erickson, "How China Got There: Beijing's Unique Path to ASBM Development and Deployment," *China Brief* 13.12 (June 7, 2013). *http://www.jamestown.org/programs/chinabrief/single/?tx_ttnews per cent5btt_news per cent5d=40994&tx_ttnews per cent5bbackPid per cent5d=25&cHash=913594fd5785c82b653cc6b00de7850c#.Ueh_F5U-aX1*; U.S. Department of Defence, *Annual Report to Congress on China's Military Power, Military and Security Developments Involving the People's Republic of China 2013* (Washington, DC: 2013), pp. 5-6.

11   Ronald O'Rourke, *China Naval Modernisation: Implications for U.S. Navy Capabilities – Background and Issues for Congress* (Washington, DC: Congressional Research Service, July 5, 2013), pp. 3-38; U.S. Department of Defence, *Annual Report to Congress on China's Military Power, Military and Security Developments Involving the People's Republic of China 2013* (Washington, DC: 2013), p. 42; U.S. Office of Naval Intelligence, *The People's Liberation Army: A Modern Navy with Chinese Characteristics* (Suitland, MD: 2009), pp. 18-28; and Anthony H. Cordesman and Martin Kleiber, *Chinese Military Modernisation: Force Development and Strategic Capabilities* (Washington, DC: Center for Strategic and International Studies, April 2007), p. 132.

12   U.S. Department of Defence, *Annual Report to Congress on China's Military Power, Military and Security Developments Involving the People's Republic of China 2013* (Washington, DC: May 2013), pp. 6-7; Michael Cole, "China's Growing Long-Range Strike Capability," *Diplomat*, August 13, 2012. *http://thediplomat.com/flashpoints-blog/2012/08/13/chinas-growing-long-range-strike-capability/*.

13   U.S. Department of Defence, *Annual Report to Congress on China's Military Power, Military and Security Developments Involving the People's Republic of China 2013* (Washington, DC: 2013), p. 31.

U.S. Office of Naval Intelligence, *The People's Liberation Army: A Modern Navy with Chinese Characteristics* (Suitland, MD: 2009), p. 23.

14   U.S. Department of Defence, *Annual Report to Congress on China's Military Power, Military and Security Developments Involving the People's Republic of China 2013* (Washington, DC: 2013), pp. 10, 31.

15   Ibid, p. 6.

16   Ibid

17   Bernard Cole, "The Future Chinese Carrier Force," *U.S. Naval Institute*

News, May 8, 2013. *http://news.usni.org/2013/05/08/the-future-chinese-carrier-force*

18   U.S. Department of Defence, *Annual Report to Congress on China's Military Power, Military and Security Developments Involving the People's Republic of China 2013* (Washington, DC: May 2013), p. 6; Xinhua, "China refitting aircraft carrier body for research, training," July 27, 2011. *http://eng.mod.gov.cn/TopNews/2011-07/27/content_4284108.htm.*

19   U.S. Department of Defence, *Annual Report to Congress on China's Military Power, Military and Security Developments Involving the People's Republic of China 2013* (Washington, DC: 2013), p. 6; Agence France-Presse, "China To Build 2 More Aircraft Carriers: Taiwan," DefenceNews.com, May 21, 2012.

20   Andrew S. Erickson and Gabe Collins, "China Carrier Demon Module Highlights Surging Navy," *National Interest*, August 6, 2013. *http://nationalinterest.org/commentary/china-carrier-demo-module-highlights-surging-navy-8842*; Peter W. Singer and Jeffrey Lin, "About those Chinese Aircraft Carrier Pics: What We Know and What We Can Guess," *Defence One*, August 5, 2013. *http://www.defenceone.com/technology/2013/08/those-chinese-aircraft-carrier-pics-what-we-know-what-we-can-guess-and-what-we-cant/68114/.*

21   U.S. Department of Defence, *Annual Report to Congress on China's Military Power, Military and Security Developments Involving the People's Republic of China 2013* (Washington, DC: 2013), p. 6.

22   Andrew Erickson and Gabe Collins, "China Carrier Demo Module Highlights Surging Navy," *The National Interest*, August 6, 2013. *http://nationalinterest.org/commentary/china-carrier-demo-module-highlights-surging-navy-8842*; and U.S. Department of Defence, *Annual Report to Congress on China's Military Power, Military and Security Developments Involving the People's Republic of China 2013* (Washington, DC: 2013), pp. 5-7.

23   BBC News, "China 'Buys Fighter Jets and Submarines from Russia,'" March 25, 2013. *http://www.bbc.co.uk/news/world-asia-21930280*; Robert Foster, "Russia to Sell, Co-Produce Lada-class Submarines to China," December 20, 2012. *http://www.janes.com/article/19682/russia-to-sell-co-produce-lada-class-submarines-to-china.*

24   Ronald O'Rourke, *China Naval Modernisation: Implications for U.S. Navy Capabilities – Background and Issues for Congress* (Washington, DC: Congressional Research Service, July 5, 2013), pp. 3-38; U.S. Department of Defence, *Annual Report to Congress on China's Military Power, Military and Security Developments Involving the People's Republic of China 2013* (Washington, DC: 2013), pp. 7-8, 39.

25   David Lague, "Chinese transport 'workhorses' extending military's reach,"

Reuters, February 25, 2013. *http://www.reuters.com/article/2013/02/25/us-china-military-ships-idUSBRE91O15V20130225*; David Axe, "This Simple Ship Could Let the Chinese Navy Circle the Globe," *Wired*, January 16, 2013. *http://www.wired.com/dangerroom/2013/01/china-new-oiler/*; and Christopher D. Yung and Ross Rustici, *China's Out of Area Naval Operations* (Washington, DC, National Defence University Institute for National Strategic Studies: 2010).

26   Andrew Erickson and Gabe Collins, "China Carrier Demo Module Highlights Surging Navy," *The National Interest*, August 6, 2013. *http://nationalinterest. org/commentary/china-carrier-demo-module-highlights-surging-navy-8842*.

27   Data extracted from IISS, The Military Balance published from 2007 onwards.

28   The Military Balance 2013, IISS, Routledge, London, 201, pp. 293-294.

29   Ibid.

30   Ibid.

31   Ibid.

32   Harry Kazianis, Did China Test its "Carrier-Killer?", The Diplomat, January 24, 2013, accessed from http://thediplomat.com/2013/01/did-china-test-its-carrier-killer/ on May 13, 2013.

5

PLA Second Artillery Force (PLASAF) Modernisation

"A world without nuclear weapons would be less stable and more dangerous for all of us"

- Margaret Thatcher

The Defence White Paper "The Diversified Employment of China's Armed Forces 2013" states that the PLA Second Artillery Force is a core force for China's strategic deterrence. It is mainly composed of nuclear and conventional missile forces and operational support units, primarily responsible for deterring other countries from using nuclear weapons against China, and carrying out nuclear counterattacks and precision strikes with conventional missiles."[1] The white paper also lays down the following trends for modernisation of the PLA SAF[2]. These are:

- The the principle of building a lean and effective force will be the essence of all future force development.

- The goal of fighting in the conditions of informatalization will be pursued.

- The reliance will be on indegenized R&D and innovations in all fields of weaponry and equipment development.

- Modernisation of the current equipment will be done by using modern, indegenous and proven technology, enhancing the safety, reliability and effectiveness of its missiles, improving its force structure of having both nuclear and conventional missiles, strengthening its rapid reaction, effective penetration, precision strike, damage infliction, protection and survivability capabilities.

- The PLASAF capabilities of strategic deterrence, nuclear counterattack and conventional precision strike will be further enhanced.

Also as per the White Paper the PLASAF has under its command missile bases, training bases, specialized support units, academies and research institutions. Its present holding of missile include a series of "Dong Feng" ballistic missiles and "Chang Jian" cruise missiles.

As per the US Annual Report to Congress, Military and Security Developments Involving the People's Republic of China 2014, The PLASAF is "developing and testing several new classes and variants of offensive missiles, forming additional missile units, upgrading older missile systems, and developing methods to counter ballistic missile defences."[3]

From the above the following trends can be derived:

- The PLASAF is working towards increasing the overall number of missiles in its inventory.

- The missiles in the future will be solid fuel propelled, a shift from the present day mix of solid and liquid fuel mix.

- The survivablity of the missiles of all catogories will be enhanced by adding mobile delivery systems with higher survivablity. The Annual Report to Congress of US Department of Defence also refers to this fact and states "The Second Artillery continues to modernize its nuclear forces by enhancing its silo based inter continental ballastic missiles (ICBMs) and adding more more survivable mobile delivery systems".[4]

- The new and indegenious missile systems being introduced in to the PLA SAF will have re loadable launchers.

- There will a shift from a purely nuclear missile force towards a mixed nuclear and conventional force.

- In consonance with the doctrine of Local Wars under Conditions of Informatization, focus will be on building a reliable, mobile and accurate short range missile force.

The Command and Control Structure of the PLA SAF[5]

Table 5.1: Command and Control Structure PLA SAF

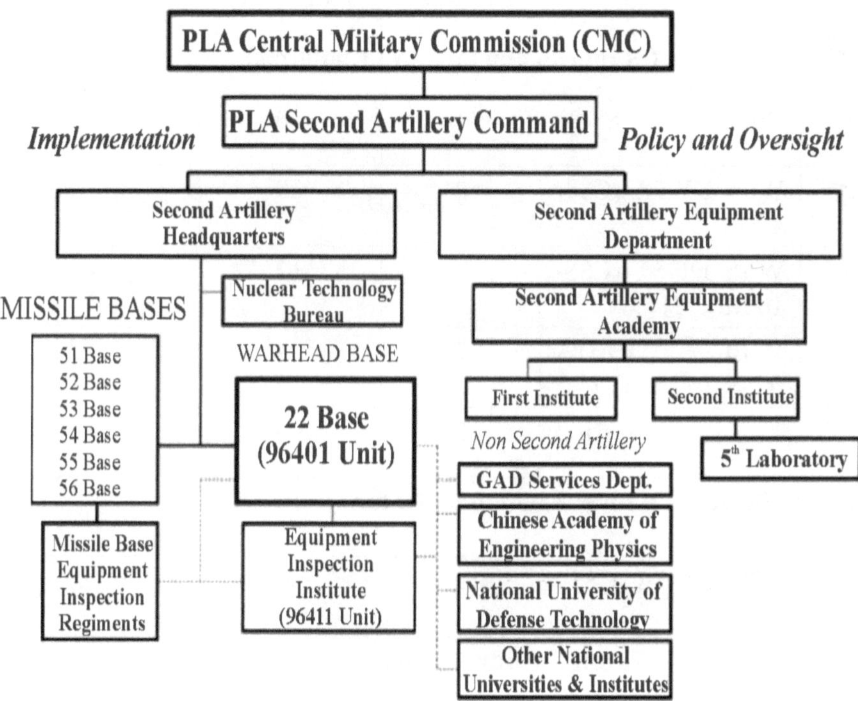

Source Project 2049 Institute

According to The Military Balance 2014 released by IISS in Feburary 2014, PLA SAF has seven ICBM brigades, one IRBM brigade, 11 MRBM brigades, eight SRBM brigades, two SSM brigades and two SSM training brigades[6]. The details of these are as under[7];

Table 5.2: Details PLA SAF Brigades

Ser No	Type of Missile	Number of Brigades	Number of Missile Launchers	Remarks
ICBM Brigades (Range: 5,400 kms to 13,000kms)				
1.	With DF-4 (CSS-3)	1	10	
2.	With DF-5A (CSS-4 Mod 2)	3	20	
3.	With DF-31 (CSS-10 Mod 1)	1	12	
4.	With DF-31A (CSS-10 Mod 2)	2	24	
	Total	7	66	
IRBM Brigade (Range: 3,000 to 5,400 kms)				
5.	With DF-3A (CSS-2 Mod)	1	6	
MRBM Brigades (Range: Above 1750 kms)				
5.	With DF-16 (CSS-11) (reported)	1	12	The DF-16 MRBM is in service but its not clear as to which formation it has been assigned to.
6.	With DF-21 (CSS-5 Mod 1)	1	80	
7.	With DF-21A (CSS-5 Mod 2)	6		
8.	With DF-21C (CSS-5 Mod 3)	2	36	

Ser No	Type of Missile	Number of Brigades	Number of Missile Launchers	Remarks
9.	With DF-21D (CSS-5 Mod 4 -AShBM) (reported)	1	6	
	Total	11	134	
SRBM Brigades (Range: 1000 kms or less)				
10.	With DF-11A/ M-11A (CSS-7 Mod 2)	4	108	
11.	With DF-15/ M-9 (CSS-6)	4	144	
	Total	8	252	
SSM Brigades				
12.	With DH-10 (LACM)	2	54	
13.	Training	2		

Deployment of the Missile Base Level Storage[8]

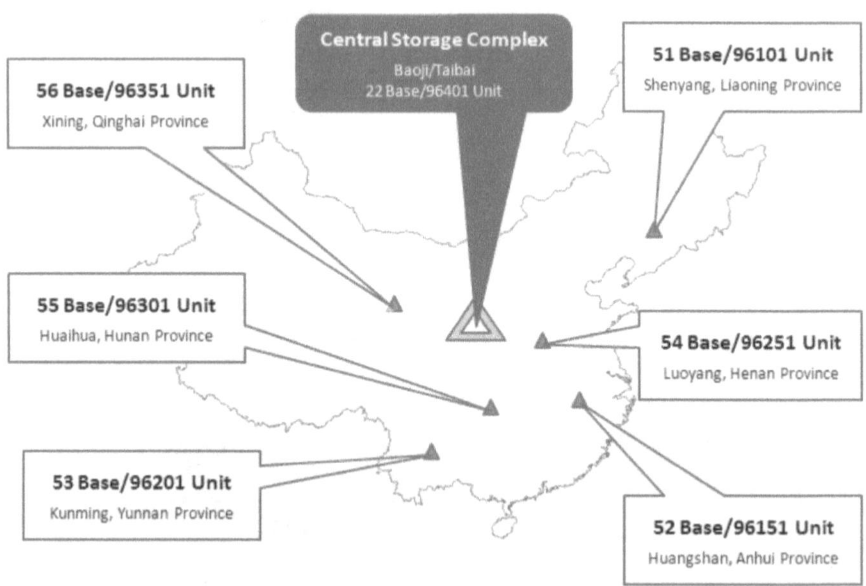

Details: PLASAF Missile Systems and Launchers

Table 5.3: PLA SAF Missile Systems and Launchers

Missile System	Fuel	Launch Mode	Reusable Launcher	Missiles per launcher
DF-3/ CSS-2	Liquid	Road mobile erector	?	?
DF-4/ CSS-3	Liquid	Silo elevate-to-launch; cave based roll out to launch TEL	No ?	1?
DF-5/ CSS-4	Liquid	Silo elevate-to-launch; cave based roll out to launch TEL	No ?	1?
DF-15/ CSS-6/ M-9	Solid	Road Mobile TEL	Yes	6

Missile System	Fuel	Launch Mode	Reusable Launcher	Missiles per launcher
DF-11/ CSS-7/ M-11	Solid	Road Mobile TEL	Yes	6
DF-21/ CSS-5	Solid	Road Mobile TEL	Yes ?	?
DF-31/ CSS-9	Solid	Road Mobile TEL	Yes ?	?

Organisation of PLA SAF Brigade.   According to Mark A. Stokes a PLA SAF brigade has six launch battalions with two companies each (a "6/2" structure). Each company likely has a launch platform (either silo or mobile launcher) and associated support vehicles in its table of organization and equipment, and stores the equipment in battalion garrison facilities. Therefore, based on each brigade's table of organization and equipment, it may be assigned at least 12 launch platforms. Other battalions within a brigade are responsible for missile diagnostics, check out, warhead mating, and other functions, usually in an underground facility (referred to as a "central depot") operated by the brigade's site management battalion. As many as six subordinate companies under a site management battalion oversee missile-related preparation, pre-surveyed launch sites, storage, and other facilities. Among site management battalion responsibilities include underground facility management such as power and electricity, water, air conditioning, and ventilation. A service battalion is responsible for security and concealment, camouflage, and deception[9].

On analysing the number of brigades in terms of percentages of the tolat number of brigades and corelate it with the percentages of missile launchers held the following emerge:

Table 5.4: Analysis of PLA SAF Brigades and Missile Launchers

	Percentageof brigade's out of total number of brigades	**Percentage of missile launchers out of total number of missile launchers**
ICBM	24.13 per cent	12.9 per cent
IRBM	3.45 per cent	1.17 per cent
MRBM	37.83 per cent	26.17 per cent
SRBM	27.58 per cent	49.21 per cent
SSM	6.89 per cent	10.54 per cent

Following can be inferred from the above:

- SRBM brigades are about 27.5 per cent of the total missile brigades, they hold about half the total number of launchers. Suffice to say that in order to have more launchers, new formations will have to be raised. Another inference can be that the PLA SAF has, at present, no intention of expanding its SRBM capabilities and in any regional scenario that develops it will be able to manage with this force structure and capability.

- The organization of the PLASAF brigades need not be a 6/2 structure but as earlier reported may have a 3/3 structure, having 27 launchers per brigade or a 4/4 structure, having 16 launchers per brigade[10].

- However in case of the ICBMs, IRBMs, MRBMs and SSMs the percentage of the missile brigades is higher than the percentage of the missile launchers, suggesting that the PLASAF has the organizations for ICBMs, IRBMs, MRBMs and SSMs, where the existing brigades can absorb additional missiles and launchers without any increment in the number of formations.

Trends PLA SAF Missile Holdings

Table 5.5: Trends PLA SAF Missile Holdings

Type of Missile	2014[11]	2013[12]	2011[13]	2010[14]	2009[15]	2005[16]	1992[17]
ICBM	**66**	**72**	**66**	**66**	**46**	**30+**	**8**
• DF-4 (CSS-3)	10	10	10	10	20		2
• DF-5 (CSS-4)							6
• DF-5A (CSS-4 Mod 2)	20	20	20	20	20	24	
• DF-31 (CSS-9)	-	12	12	12	6	8	
• DF-31A (CSS-9 Mod 2)	-	30	24	24			
• DF-31 (CSS-10 Mod 1)	12		-				
• DF-31A (CSS-10 Mod 2)	24		-				
• DF-41	Under development. Likely to have MIRV capability						
IRBM	**6**	**2**	**118**	**118**	**35**	**110+**	**60**
• DF-3 (CSS-2							60

Type of Missile	2014[11]	2013[12]	2011[13]	2010[14]	2009[15]	2005[16]	1992[17]
• DF-4 (CSS-3)						20	
• DF -3A (CSS-2 Mod)	6	2	2	2	2	32	
• DF -21 (CSS-5)			80	80	33	60	
• DF -21C (CSS-5 Mod 3)			36	36			
MRBM	**134**	**122**					
• DF -16 (CSS-11)	12	Some					
• DF -21/ DF-21A (CSS-5 Mod 1/2)	80	80					
• DF - 21C (CSS-5 Mod 3)	36	36					
• DF-21D (CSS-5 Mod – 4 ASBM)	Reported	Reported					
SRBM	**252**	**252**	**204**	**204**	**725**	**86**	
• DF-11A/ M-11A (CSS-7 Mod 2)	108	108	108	108	500	32	
• DF-15/ M-11 (CSS-6)	144	144	96	96	225	24	

Type of Missile	2014[11]	2013[12]	2011[13]	2010[14]	2009[15]	2005[16]	1992[17]
• DF-7 (CSS-8)						30	
LACM: 54	54	54	54	54			
• DH-10	54	54	54	54			

From the analysis of the above facts several key indicators of a shift from a medium-/intermediate-range nuclear force to a bifurcated multi-mission force can be derived.

- The first such indicator is the diminishing number of missile launchers solely suited to nuclear missions. Even if the DF-21C/D missile systems are ignored and one counts the DF-21 series as a nuclear-only class, the percentage of the SAF's missile launcher strength suited only for nuclear mission's drops from 100 per cent in 1992 to slightly over 40 per cent in 2012. Roughly 60 per cent of the current SAF arsenal can conduct effective conventional missions and thus contribute to victory in non-nuclear Local Wars under Conditions of Informatization. The reason for this significant change is the introduction of precision or near- precision strike SRBMs and LACMs. When SRBMs first appeared on the graph in 2000, they accounted for 30 per cent of the SAF's missile launchers; by 2013, SRBMs accounted for approximately 50 per cent. This change is complemented by the introduction of cruise missiles: by 2011, LACMs accounted for roughly 11 per cent of SAF strength. These trends occur in contrast to the effective destruction of the SAF's nuclear intermediate-range ballistic missile (IRBM) force. In 1985, the SAF's nuclear IRBMs accounted for over 50 per cent of the force; by 2013, the total was roughly 0.4 per cent.

- The second major indicator of a shift in SAF doctrine and capability is the significant growth in the relative size of the ICBM arsenal. Not only does the ICBM force increase in relative size from 5 per cent to 14 per cent, but also much of the growth is due to modern DF-31 and DF-31A ICBMs. This capability will be further enhanced once DF-41 is also operationalized. This trend may be an indication of a shifting priority from regional and Eurasian deterrence missions to intercontinental deterrence missions. Consequently, not only have the SAF's equipment holdings revealed a shift from nuclear to nuclear and conventional missions, it is possible that the same equipment holdings also indicate a shift in the priority of nuclear deterrence missions.

- The third indicator is the change in the geographic range of the force. In 1985, 100 per cent of the SAF's missile force could reach up to the second island chain. In 2012, the composition of the SAF is such that only roughly 15 per cent of the SAF's capabilities can hit up to the second island chain or the US base on Guam. This change indicates a significant shift in priorities from the second island chain and beyond to China's immediate periphery. Such a shift is fully in line with the Local Wars concept.

PLASAF in 2025

- Today China has the largest conventional missile arsenal in entire Asia- Pacific. The force development of the PLASAF will follow the trend of developing a conventional and nuclear capable missile force with a focus to continue to develop conventional missile forces with increased precision or near precision especially of the MRBMs, SRBMs and LACMs.

- As far as the ICBMs are concerned two shifts that are likely to emerge are:

  ➢ The first shift that is likely to emerge is that the employment of ICBMs from a purely nuclear to a mix of nuclear and conventional missions, thereby creating a dilemma in the mind of the adversary and over burdening the ABM defences.

  ➢ The priority of development of ICBMs will be for nuclear deterrence missions. For this the emphasis will be on the development of solid fueled road and rail mobile ICBM systems rather than the silo based. Also the SLBM capability will also witness a push and effort of the PLASAF and the PLAN will be early development and operationalization of the Jl-3 SLBMs. In addition to this the effort of operationalizing the DF-41 ICBM with MIRV will be enhanced with increase in the numbers of DF-41, DF-31 and DF-31A.

- The PLASAF missile force will move towards becoming a multi mission missile force i.e. to say that the missile launchers will become increasingly capable of launching nuclear and conventional missiles and there will be diminishing numbers of missile launchers only for nuclear missions

- The PLASAF will continue to move in a direction to develop an arsenal which will give it a capability to conduct effective conventional missions in addition to nuclear deterrence missions. Accordingly the PLASAF will focus on a precision or near precision capability for its SRBMs/ LACMs and accordingly there will be a sharp decrease in numbers of IRBMs presently held by the PLASAF.

- What do these trends indicate? From the above trends on can postulate that the PLASAF is aiming at shorter ranges, mix of conventional and nuclear missiles with multipurpose launchers: hence the focus on the immediate neighborhood.

Endnotes

1   Information Office of the State Council, *"The Diversified Employment of China's Armed Forces"*, April 2013, Beijing.

2   Ibid.

3   U.S. Department of Defence, Annual Report to Congress, *"China's Military Power, Military and Security Developments Involving the People's Republic of China 2014"*, p. 6, Washington, USA.

4   Ibid, p. 7.

5   Ian Easton & Mark Stokes, Half Lives: A Preliminary Assessment of China's Nuclear Warhead Life Extension and Safety Program, Project 2049, July 29, 2013, p. 5.

6   The International Institute of Strategic Studies, *The Military Balance 2014"*, February 2014, p. 231, London.

7   Ibid.

8   Mark A. Stokes, China's Nuclear Warhead Storage and Handling System, Project 2049, March 12, 2010, p. 7.

9   Stokes, Mark A., *"Hearing on Developments in China's Cyber and Nuclear Capabilities"*, Prepared Statement before The US-China Economic and Security Review Commission, March 26, 2012, p. 3, Virginia, USA.

10   Allen Kenneth and Kivlehan-Wise Maryanne, *"Implementing PLA Second Artillery Reforms"*, China's Revolution in Doctrinal Affairs: Emerging Trends in the Operational Art of the Chinese People's Liberation Army, p. 177.

11   The International Institute of Strategic Studies, *The Military Balance 2014"*, February 2014, p. 231, London.

12   The International Institute of Strategic Studies, *The Military Balance 2013"*, March 2013, p. 287, London.

13   The International Institute of Strategic Studies, *The Military Balance 2011"*, March 2011, p. 230, London.

14   The International Institute of Strategic Studies, *The Military Balance 2010"*, February 2010, p. 399, London.

15   The International Institute of Strategic Studies, *The Military Balance 2009"*, February 2010, p. 382, London.

16   The International Institute of Strategic Studies, *The Military Balance 2004-5"*, February 2005, p. 170, London.

17   The International Institute of Strategic Studies, *The Military Balance 1991-92"*, 1991, p. 150, London.

6

Dragon Fire and Indian Response

"The conqueror is always a lover of peace; he would prefer to take over our country unopposed"

- Karl von Clausewitz

Chairman Mao, in December 1936, said 'Unquestionably, victory or defeat in war is determined mainly by the military, political, economic and natural conditions on both sides, however, not by these alone'. It is also determined by each side's subjective ability in directing the war. In his endeavor to win a war, a military strategist cannot overstep the limitations imposed by the material conditions; within these limitations, however, he can and must strive for victory. The stage of action for a military strategist is built upon objective material conditions, but on that stage, he can direct the performance of many a drama, full of sound and color, power and grandeur'.[1] Almost 77 years later we see an almost complete manifestation of this as part of PRCs coercive or otherwise military diplomacy and assertive engagements.

Chinese military modernization and force development has been an ongoing process for almost now two decades. Today PRC propagates the idea of a 'Peaceful Rise', of which a modern military is an important component. The strategy has also has seen a paradigm shift from the Mao's Peoples War to Deng's 24 Character Strategy to today's Xi's China Dream (*Zhongguo meng)* this too in the backdrop of the 'Three Warfare Doctrine'. At a meeting of South-East Asian nations in 2010, China's foreign minister Yang Jiechi, facing a barrage of complaints about his country's behavior in the region, blurted out the sort of thing polite leaders usually prefer to leave unsaid. "China is a big country," he pointed out, "and other countries are small countries and that is just a fact."[2]

In this backdrop the likely response of India need to be worked out.

The fact is that China is going to annex Tibet and geneses of the Sino – Indian border war of 1962 were in that time, well known to Indian leadership and the decision makers. Today 'Rise of China' is an accepted fact in the international order and this rise includes a rapidly modernizing military an equal acceptability.

China's Threat Perceptions

Internal Threats. Internally, the greatest threat that the Communist Party of China (CPC), the ruling dispensation fears is the loss of control of the Party. This fact gets further gets highlighted when we see that 85.13 million[3] members of the CPC rule over a population of 1.39 billion[4] Chinese. To perpetuate its rule (which China's top leaders truly believe is essential for the well-being of the country), the Chinese Communist Party has constructed a grand narrative that is founded on three critical claims: Only the Communist Party can continue to improve citizen's standard of living (and ameliorate severe social and economic disparities); only the party can maintain a stable, unified country and construct a happy, harmonious society; and only the party can effect the "rejuvenation of the Chinese nation," which stresses a firm command of "core interests" (i.e., sovereignty and territoriality) and increasing global respect.[5] In continuation, on taking over as the President of People's Republic of China Xi gave out the idea of the "China Dream", which talks of 'Two 100's': China becoming a "moderately well-off society" by about 2020, the 100th anniversary of the Chinese Communist Party, and the modernization goal of China becoming a fully developed nation by about 2049, the 100th anniversary of the People's Republic. The China Dream has three mile stones: doubling of the per capita GDP by 2021, urbanization target of one billion of her population and by 2050 China firstly regaining its position as a world leader in science and technology, in economics and business, secondly the resurgence of Chinese civilization, culture and military might and thirdly China participating actively in all areas of human endeavor. The core of this approach is economic well being of the population with strict state control on all facets of life of a Chinese citizen. This is a challenge and has inherent contradictions which is that the future Chinese leadership will face in balancing the need of an economically well off population for an inclusive and transparent government, global integration and connect, free media, private ownership individual rights and social security – which could lead to a loosening of the party rule or a partial democracy or in an extreme case complete change in the system of governance. In this kind of evolving scenario the role of and the resistance by the PLA will play a very major and dominant role both at the ideological level and in application of force.

The second major threat that the CPC faces is 'Extremism, Separatism and Terrorism' in her periphery. The violence that has been seen recently in Kunming, Beijing, Changsha and Urumqi have been termed as a violent terrorist acts allegedly by East Turkistan Separatist Movement (ETIM), a separatist force, using extreme measures. A thought comes to mind is China prepared to tackle this threat? Will this separatist movement take an Islamic turn? Or is it already an Islamic terrorist organization? How does it plan to tackle this and will plan to tackle this will involve a larger game plan to sort out issues embroiling the entire periphery? PLA has an in-depth understanding of the Islamic terrorism purported by the likes of Taliban and Al Qaida as during *Op Cyclone*, the CIA funded training of the Mujahedeen in Afghanistan, the PLA played an important role. This cooperation continued and with the relations PRC has with Pakistan, the PLA exercises enough leverage to adequately counter this threat in pure military terms. More over PLA is transforming its forces in the Chengdu and Lanzhou MAC to fight the developing counter insurgency or emerging terrorist threats. Today, in addition to the SF battalions in the Combined Corps, there is one SF battalion each in the Xinjiang and Xizang Military Districts. An inference that may be a possibility is that the PLA is waiting for the Uyghur movement to develop into a full blown insurgency, which can be labeled by the PLA as an extension of 'Islamic terrorism East wards' and then respond to it encompassing the provinces of Xinjiang and Xizang together. This response will have an international legitimacy and the separatist issues in the two provinces will be together. This inference gains some amount of legitimacy due to the fact that today the PLA has a clear cut strategy to tackle the Uyghur insurgency, they are grappling with way to tackle the issue of self immolations in Xizang (Tibet) and at present they do not have a solution. Employing strong arm tactics of excessive use of force does not only gels with international perceptions but also is seen by the Chinese leadership as the right kind of message being sent by a rising great power to the world. In addition there are other threats like the demographic inversion, economic slowdown etc, which the CPC seems capable of handling.

External Threats. China perceives no major threat externally. Today she is engaged in claiming area's perceived to be hers historically both along continental and maritime borders. The upward trend of her military modernization has ensured execution of an effective Counter Intervention strategy blunting US A2AD strategy. The assertiveness in form of salami slicing that is being expedited by the Chinese today will continue and all events will be short of an all out act of war. The issue in hand is till where will this assertiveness continue; up to the first island chain or the second island chain and beyond. This is totally dependent on the world powers how they accept the continued assertiveness and react to it.

India China Relations

Chinese Ambassador to India, Ambassador Wei Wei, in February 2014, said that the boundary question was among the three sensitive issues as far as India – China relations are concerned. The other two are the issue of trans-border rivers and China-Pakistan relations[6]. He suggested "While divergences and differences are unavoidable, most important is how we handle them. There have been accidental standoffs, but we have maintained peace and tranquility in border areas and had no single firing accident for long. In future, we should continue, based on the principle of constructive cooperation, to work together to safeguard peace and tranquility in border areas"[7]. The Indian approach is guided by simultaneously pursuing both a robust diplomatic strategy aimed at encouraging peaceful resolution of border disputes and forging strong trade and economic ties and an ambitious military modernization campaign that will build Indian air, naval, and missile capabilities. As far as the Trans Border Rivers issue is concerned, both sides have been moving forward in comparison to the other two issues. In addition to India, Pakistan in West and Bangladesh in East are also lower riparian states there by indicating that the negotiation reaching the conclusion in times ahead seems a highly plausible outcome. The other two issues are both complex and tricky.

Border Dispute

A question is asked a number of times as to what is the way ahead for resolution of the border dispute. The disputed areas between India and China are basically two; The Western Sector or the Aksai Chin, basically the area between the Kunlun Mountains and the Karakoram Range and the Eastern sector encompassing the entire state of Arunachal Pradesh. The present status of Aksai Chin is that it is occupied by the Chinese where they have developed important military infrastructure and the China National Highway (G 219), connecting Yecheng in Xinjiang to Lhatse in Tibet (2086 kms) passes through this area. This is an all weather black top road running parallel to the Line of Actual Control. A number of laterals emanate and connect the forward Chinese posts on the LAC. The construction of this highway was started in 1951 and was completed in 1957. More over today this area is of immense strategic value to the Chinese as they have invested enormous resources in this area. North of Aksai Chin is the starting of the Karakoram Highway which further connects to the Kashgar Gwadar Economic Corridor. Once operational this corridor will reduce the travel time of commodities from Middle East to Eastern parts of China to about 48 to 72 hrs from the present 20 to 25 days. Let's also not forget that this was one of the routes used by the PLA to occupy Tibet as it was an all weather road and it avoided the rebel strongholds of Kham and Amdo. The overall importance of Aksai Chin can also be ascertained from the fact that the fact that three different instances; China has offered to barter NEFA (erstwhile Arunachal Pradesh) for Aksai Chin. The first time was by Premier Zhou when he hinted at such an arrangement in 1960 and this was again hinted in the declaration of 24 October, 1962, after the first phase of the offensive. The third time it was raised was immediately after the ceasefire when China hinted that while she was ready to make concessions on the McMahon Line, but wanted to retain what she claimed in Aksai Chin.

In the Eastern Sector McMohan line demarcates the boundary between India and Tibet. This line was agreed to by Britain and Tibet in 1913 at Shimla. In addition to the British and the Tibetan representative, a Chinese representative was also present, though he did not sign the accord. The Shimla Accord was signed on 24 -25 March 1913 by the British and Tibetan representatives as a bilateral accord. Here lay the seeds of dispute. As independent India, this was an inheritance, which was not recognized by China and called Arunachal Pradesh as South Tibet[8]. On January 20, 1972, North-East Frontier Agency (NEFA) as it was called since 1955 was renamed as Union Territory of Arunachal Pradesh, which attained the status of an Indian state on February 20, 1987. The Chinese continue to contest this and express their irritation in many forms. For example the issue of stapled visa, that has been an irritant as the Chinese feel that if the visa is embossed on the passport, it may be construed that China accepts Indian sovereignty over Arunachal Pradesh.

In this existing scenario is a settlement of the border dispute possible. As Professor Srikanth Kondapalli[9], points to China's history of seeking solutions to border disputes "The Chinese resolve problems only with small or weak countries". In the 1960s, Beijing moved to resolve problems with Nepal, Afghanistan, Pakistan, Myanmar and Mongolia. But never with Vietnam and India, and did so with the Soviet Union only after it broke up in the 1990s and was considerably weakened from it. "Right now both India and China are on an upswing. By waiting to resolve the issue, China would know which way India is headed," he says. "Will India become weaker or stronger? At the same time, through massive infrastructural development, China would integrate these areas into its mainland so that any major territorial concession would not be possible." As per ex Ambassador Kanwal Sibal[10] "By not settling the border dispute, China is able to put pressure on us as well as limit our regional role. If they settle the border, our relations would improve and their choices would get limited. They will then have to do things in consultation with us." The Chinese, he says, want to keep India on the "wrong foot" as a justification to extend their influence in the neighborhood. He also feels that "China is insecure about Tibet. It feels India has a card there and if it settles the border, it would have to inevitably settle with the Dalai Lama. The two cannot be separated in their minds." Beijing, it is believed, wants to wait out until the Dalai Lama, 77 years, passes away, at which point it would install a Dalai Lama of its choosing, divvying up the Tibetans. As an outcome of the thought process and under China's pressure, Nepal has already clamped down on Tibetan exiles within its borders forcing them to roll back protests[11]. Analysts say Beijing wants to keep its claims on Tawang alive so that it can twist India's arms to crack down on Tibetan protests here.

Is economy the driver of change and push the two dominant powers of Asia towards a peaceful resolution of the border issue. As per the OECD study, Asia 2050; Realizing the Asian centaury, by doing an objective analysis of long-term economic trends, by around 2030 Asia will be the world's powerhouse just as it was prior to 1800. Currently the OECD has two-third of global output compared to one-fourth in China and India, and by 2060 these two countries will have a little less than half of world GDP with OECD's share shrinking to one-quarter. China is expected to surpass the US by 2016 to become the largest economy in the world, and India's GDP is expected to exceed that of the US by 2060. India's GDP will increase from 11 per cent to 18 per cent as a share of global GDP while China's share will remain at 28 per cent during this period, just as it has been since the dawn of civilization[12].

The primary factor holding India back will be the growing demand for food, water and energy, which is expected to double, and will require reshaping a global system that served the natural resource and security needs of 20 per cent of the population to one that will share prosperity and peace with all of humanity in an interdependent world faced with ecological limits[13].

What can be the way forward in resolution of the border dispute? Is it a dispute? The areas under dispute today are well under control of India and China. One can argue that Aksai Chin was occupied by China post Indian independence where as Arunachal Pradesh was the part of the country when we became independent. These issues get complicated when the Chinese interests in Pakistan's Northern Area's or Gilgit Baltistan are super imposed. The infrastructure developments in Pakistan's Northern Areas are very important to the Chinese as they see this not only as a land connect to the warm waters of the Arabian sea but also views it as an effective mechanism to control of Islamic terrorism into its restive periphery.

In order to overcome the complicated issues that have dogged India and China over the years, the two countries will have to reconstruct international relations theory, as the focus of both realists and idealists is on material force and material benefit whereas we now need a global vision of sharing natural resources and technology and preserving the environment. A shared vision of prosperity for four billion people who have yet to benefit from globalization will provide the legitimacy to reshape the global order.[14]

China – Pakistan Relations

China and Pakistan are two sovereign countries and as per existing international norms have close and warm relations. The issues come up when we see China – Pakistan relations in relation to India – Pakistan relations. The resolution of Kashmir issue in Pakistan's favour as a 'revenge' for creation of Bangladesh, construed as dismemberment of Pakistan is the basic premise on which the whole idea of India – Pakistan relations gravitate and the thought embedded in the minds of the Pakistani leadership and Pakistan Army. This thought propagates the support to extremism, separatism and terrorism in India and obviates any forward movement in economic ties, people to people contact, opening of borders or any other step taken in forward direction by the two countries. In this setting Pakistan see's China as an all weather friend and extends all possible help to China in furtherance of Chinese interests, in return extracting requisite assistance to adequately balance India, thus meeting requirements of both China and self. Diplomatic relations between Pakistan and China were established on 21 May 1951, which initially ambivalent towards the idea of a Communist country on its borders, Pakistan hoped that China would serve as a counterweight to Indian influence. Over the years this has been proven and the Chinese have provided the Pakistan establishment the requisite help, technology and the hard ware that makes Pakistan a nuclear weapon state with sophisticated missile technology. Similar examples are there in the diplomatic and economic fields, the Kashgar – Gwadar Economic Corridor, an $ 18 Billion investment, being the latest. In May 2011, during the visit of Prime Minister Yusuf Raza Gilani to China to celebrate 60 years of Sino – Pakistan friendship, Pakistan's ambassador to Beijing, Masood Kahn, described the robust friendship between the two countries. "We say it is higher than the mountains, deeper than the oceans, stronger than steel, dearer than eyesight, sweeter than honey, and so on."[15] True to this fact today the relationship between the two countries is a geopolitical keystone for both countries. Pakistan serves as China's closest friend both in South Asia and among Islamic countries. So close that after post *Op Neptune Spear* it is suspected that China has asked Pakistan for the remains of the American modified Black Hawk, a stealth helicopter abandoned during the raid. Meanwhile, China can help counterbalance Pakistan's arch-rival, India, including in Afghanistan. In a speech to parliament on May 9, 2011 Prime Minister Yusuf Raza Gilani struck out on an odd tangent to praise China as an "all-weather friend", providing Pakistan with strength and inspiration. Not to be outdone, President Asif Zardari issued an effusive statement of his own about a friendship "not matched by any other relationship between two sovereign countries". As the Chinese Comprehensive National Power (CNP) increases, issues of contested sovereignty gain prominence and China increasingly becomes assertive, especially in Western Pacific, it will continue to need Pakistan to balance India. Pakistan will continue to have a free access to Chinese military and economic advances, moreover in the nuclear and missile technologies. We must be prepared to accept that once J – 20s are operational, Pakistan may be the first country to fly in her air force after the PLA. In light of this if we play a China – Pakistan collusive scenario, the analysis of the conventional threat as given below.

An Analysis of the Military Balance in a China – Pakistan Collusive Scenario

The detailed Military balance in a China – Pakistan scenario is as under.

Table 6.1: Analysis of Military Balance in a China – Pakistan Collusive Scenario

	China (Source: The Military Balance 2013)	India (Source: www.globalsecurity.com)		Pakistan (Source: The Military Balance 2013)	Remarks
		Against China	Against Pakistan		
Ground Forces					
Personnel (Active)	1,600,000	1,129,000*		550,000	* Source: The Military Balance 2013, IISS
Combined Corps (Group Armies)/ Corps	18	3	10	9	
Infantry Divisions	14#	11	22	18	#Mot Inf Divs
Infantry Brigades	21	2	7	5	
Mechanized Infantry Divisions	9			2	
Mechanized Infantry Brigades	12		2	1	
Armed Divisions	5		3	2	

	China (Source: The Military Balance 2013)	India (Source: www. globalsecurity. com)		Pakistan (Source: The Military Balance 2013)	Remarks
		Against China	Against Pakistan		
Armed Brigades	11		8	7	
Artillery Divisions	2		3		
Artillery Brigades	17			10	
Airborne Divisions	3				
Airborne Brigade			1		
Amphibious Divisions	2				
Amphibious Brigades	1				
Main Battle Tanks	7,430+	3,274+*		2,411	
Artillery pieces	12,367	9,682+*		4,607+	
Navy*					
Total Manpower	**255,000**	**58,350**		22,000	
Submarines	**65**	-			
Strategic: SSBN	**4**	-			
Tactical	**61**	**15**		8	
SSN	5	1			

| | China (Source: The Military Balance 2013) | India (Source: www. globalsecurity. com) | | Pakistan (Source: The Military Balance 2013) | Remarks |
		Against China	Against Pakistan		
SSK	55	14		5	
SSB	1	-			
SSI				3	
Principal Surface Combatants	**77**	**21**		10	
Aircraft Carriers. CV	1	1			
Destroyers	14	11			
DDGHM	12	6			
DDGM	2	5			
Frigates	**62**	**12**		10	
FFGHM	29	11		3	
FFGM	1	-			
FFGM	4	-		4	
FFHM				2	
FFG	28	-			
FFH	-	1		1	
Amphibious					

	China (Source: The Military Balance 2013)	India (Source: www.globalsecurity.com)		Pakistan (Source: The Military Balance 2013)	Remarks
		Against China	Against Pakistan		
Principal Amphibious Vessels: LPD	2	1			
Landing Ships	85	10			
LSM	59	5			
LST	26	5			
	151	8			
LCU	120	8			
LCM	20	-			
LCAC	1	-			
UCAC	10	-		4 (LCAC)	
Logistics and Support Ships	**205**	**50**		11	
Naval Aviation	**26,000**	**7,000**			
Air Craft (combat capable)	**341**	**34**		7	
BBR	30	-			
FTR	72	15			

	China (Source: The Military Balance 2013)	India (Source: www. globalsecurity.com)		Pakistan (Source: The Military Balance 2013)	Remarks
		Against China	Against Pakistan		
FGA	200	10			
ASW	4	9		7	
ELINT	7	-			
AEW&C	6	-			
ISR	7	-			
TKR	3	-			
TPT	66	37			
TRG	106+	12			
MP	-	14		6	
Helicopters					
ASW	44	54		12	
AEW	10	9			
SAR	6	-			
TPT	43	-			
MRH	-	53		6	
<u>**Air Force**</u>					
Total Strength	**300,000- 330,000**	**127,000**		70,000	

	China (Source: The Military Balance 2013)	India (Source: www. globalsecurity. com)		Pakistan (Source: The Military Balance 2013)	Remarks
		Against China	Against Pakistan		
Air Craft (combat capable)	**1,903**	**870**		423	
FTR	842	63		200	
FGA	543	736		174	
ATK	120	-			
EW	13	-			
ELINT	4	-		2	
ISR	51	3		10	
AEW&C	8	3		6	
C2	5	-			
TKR (AAR)	10	6		4	
TPT	326+	238		34	
TRG	950	241		143	
Helicopters					
MRH	22	226		15	
TPT	28	106		4	

	China (Source: The Military Balance 2013)	India (Source: www. globalsecurity. com)		Pakistan (Source: The Military Balance 2013)	Remarks
		Against China	Against Pakistan		
ATK	-	20		#	# Anti Tank heli-copters held with Army

An Analysis

- From the above table it is evident that the deployment of Indian Army is balanced towards our Western borders to tackle the Pakistan threat. Against China at present the deployment portrays a defensive mind set, with just adequate troops to maintain a stable defensive posture.

- Some formations are designated as Dual Task Formations, but the time of decision, movement and acclimatization needs to be factored in before their employment. More over their training and equipment profile may not lend itself for their efficient employment.

- Recent media reports say that government, post the May 2013, Dibsang incident with China, have sanctioned Rs 84,000 Cr for infrastructure development all along the Northern borders, including raising of a mountain strike corps. However the location of the mountain strike corps need to analysed in detail for its employment in a time frame that will be decisive in a conflict with China or Pakistan as the case may be.

- As far as the maritime and aerospace domains are concerned, Pakistan's capabilities on its own do not pose any challenge to India, however a combination of Chinese and Pakistani capabilities will pose a formidable threat. The severity of this threat further gets enhanced if we consider the facilities created by the Chinese in the IOR and the ongoing enhancement of infrastructure in TAR.

- The plausibility of such a scenario manifesting is in a realm of 'Not Likely', but what it brings out in terms of short comings of our present force structuring needs to be analysed in detail, corrections if any need to be mulled over and incorporated if required.

A detailed analysis of the Military Balance and equipment holdings of India and China is attached as Appendix 6.

Manifestation of Chinese Assertiveness onto India

The analysis of the present ORBAT of the MACs, transformation (both brigadization and mechanization) of the group armies and the indicated tasking of the PLAA suggests that no major shift or change in the deployment and orientation is anticipated in near future. Military and Security Developments Involving the Peoples Republic of China 2013[16], The Annual Report to Congress by the US Department of Defence classifies the Group Armies (GA) under the various Military Area Commands (MAC) as under:-

Table 6.2: Group Armies – Primary and Secondary Missions

Shenyang MAC	Nanjing MAC	Lanzhou MAC
• 16 GA: Defensive, Offensive CT • 39 GA: RRU, Offensive MF • 40 GA: Defensive, Offensive CT	• 1 GA: Amphibious, Offensive CT • 12 GA: Amphibious, Offensive CT • 31 GA: Amphibious, Offensive CT	• 47 GA: Defensive, Offensive CT • 21 GA: Offensive MF, Defensive
Beijing MAC • 65 GA: Defensive • 38 GA: RRU, Offensive MF • 27 GA; Defensive	**Guangzhou MAC** • 15 AB: RRU, Offensive MF • 41 GA: Offensive CT, Amphibious • 42 GA: Amphibious	**Abbreviations** MF – Mobile Force RRU – Rapid Reaction Force CT – Complex Terrain (mountain, urban, jungle etc)
Jinan MAC • 26 GA: Offensive CT, Defensive • 20 GA: Offensive CT, Defensive • 54 GA: Offensive MF, Amphibious	**Chengdu MAC** • 13 GA: Defensive, Offensive CT • 14 GA: Defensive, Offensive CT	

From the above it can be assessed that up to eight CCs; three Dual Tasked and five Offensive CCs can be applied in offensive against India. Details are as under:-

- **Chengdu MAC.** 13 Combined Corps and 14 Combined Corps (Dual Tasked).

- **Lanzhou MAC.** 21 Combined Corps (Offensive) and 47 Combined Corps (Dual Tasked).

- **Guangzhou MAC.** 41 Combined Corps (Offensive).

- **Jinan MAC.** 20 Combined Corps and 54 Combined Corps (Offensive).

- **Beijing MAC.** 38 Combined Corps (Offensive).

The GAs from the Shenyang MAC, owing to the distance and to cater for the contingency in Northern parts of China and the Nanjing MAC, as the primary task of the formations is amphibious and to cater for the 'Taiwan contingency, have not been taken into account. Based on these assumptions, a likely force level that may be applied against India is as under:-

Table 6.3: Likely force levels that may be applied against India in any contingency (in 2014)[17]

MAC	Inf/ Mot/ Mech Div	Inf/ Mot/ Mech Bde	Armd Div	Armd Bde
Chengdu	4 (1 Mech+3 Mot)	2 (Mtn Inf)		2
Lanzhou	4 (1 Mech+3 Mot)	5 (2 Mech + 2 Mot +1 Mot Regt)		2
Guangzhou	2 (1 Mech+1 Mot)			1
Jinan	2 (Mech)	3 (Mech)		2
Beijing	2 (Mech)		1	
Total*	**14 (7 Mech + 7 Mot)**	**10 (5 Mech + 2 Mot + 2 Mtn Inf + 1 Mot Regt)**	**1**	**7**

(* This force level includes the formations of the Xizang and Xinjiang MDs.)

Abbreviations

- Inf – Infantry

- Mot – Motorized

- Mech – Mechanized

- Armd – Armoured

- Div – Division

- Bde – Brigade

- Mtn – Mountain

The analysis of the above is as under:-

- 14 divisions (seven mechanized and seven motorized), 10 combined operational divisions/ brigades (five mechanized, two motorized, two mountain infantry brigades and one motorized infantry regiment), one armed division and seven armed combined operational divisions/ brigades cancan be applied against India. It would be logical to plan on a PLAA employment of about 32 divisions/ brigades (including armoured brigades/ divisions) for an Indian contingency in any emerging operational scenario by 2025.

- In addition, in the event of any conflict with India, the Border Defence (BD) regiments will adopt a forward posture in the vulnerable sectors or in areas of launch pads for subsequent operations. The initial offensive could be commenced by pre acclimatized forces of MAC or of CCs indirectly postured in the garb of training or for maintaining law and order or for anti terrorist operations.

- For the follow on offensive, the PLA could use the offensive or Dual Tasked CCs from within Chengdu or Lanzhou MACs. This will also entail a shift in the command and control from the MAC to the superimposed HQs of the CMC.

- Subsequent the offensive formations from other MACs particularly those in the hinterland could be utilized.

- It is pertinent to mention that in the event of an Indian contingency,

PLA will have to maintain a strong defensive and deterrence posture in the Western Pacific. Therefore own intelligence agencies would be required to undertake an in-depth analysis of the prevailing politico-security situation in the region.

- There are three amphibious CCs of Nanjing MAC and one of Guangzhou MAC- likely to be employed for amphibious operations in the South China Sea/ East China Sea and for Taiwan contingency. In the event of any military contingency in the IOR (Andaman and Nicobar) one of these amphibious divisions could be earmarked for amphibious operations.

- As far as 15 AB CC is concerned, its main employment could be for Taiwan contingency or capture of contested island territories. The possibility of one of the divisions for Indian contingency exists. However, the terrain on the continental borders would permit employment of brigade sized force only. The possibility of employment of an AB division together with an amphibious division against our island territories in IOR cannot be ruled out.

- From the Chinese strategic perspective the IOR is gaining salience in its strategic security calculus. Presently, approximately 80 per cent of China's oil flows through the region. Therefore protection of SLOCs in the IOR would assume higher priority in China's maritime security strategy. Chinese understands the limitations that the geography poses on deployment of a major naval force in the IOR. Need for enhanced naval deployment in the Western Pacific viz-a viz US and Japan would be yet another consideration in deciding the force level for deployment in the IOR. The military logic therefore suggests that the China's maritime operations will come to heavily rely on a standoff strike capability (employing AShBMs, AShCMs and DF-21Ds) and submarine fleets equipped with conventional and nuclear missiles and this would require a very high level of informationalization in space, sub space, air, surface and sub surface/ underwater with a high degree of stealth and survivability. The PLAN, however, would utilize its frigates and destroyers, in a limited manner in support of missile and submarine operations. India will therefore have to factor a viable counter capability to neutralize PLANs asymmetric advantage in missile and submarine warfare.

- Nonetheless, the developments of the South Sea Fleet (likely to be employed in the IOR) need to be closely watched, to include, the force structuring, development and nature of operational exercises being conducted and the participation of the warships of the fleet in out of area deployment.

- As far as the PLAAF is concerned, the air fields in Tibet need to be closely observed for deployment of modern four and half/ fifth generation fighters and the force multipliers to include the AEW&C aircrafts, EW aircrafts, AWACs and AAR. Also the infrastructure being developed in these airfields, to include the hangers, refueling facilities and the command and control facilities, need to be closely monitored.

- The cyber offensive and Second Arty Corps will play a key role in undertaking key point strikes against Vulnerable Areas and Vulnerable Points during all phases of campaign. This aspect needs to be fully analysed in assessing China's offensive plans and their prosecution during non contact and contact stages of a conflict. The fact that the cyber offensive does not have to wait for the hostilities to break and is a continuous ongoing campaign has to be kept in mind always by Indian policy and defence planners in formulating the planning parameters for both force development and force structuring.

- China together with Pakistan has a strong potential to fuel proxy war against India. This aspect too would merit a detailed examination and assessment in reviewing and refining our Internal Security and Rear Area Security plans.

- Finally, considering the extensive nature of communication and logistics infrastructure in Tibet, it would be reasonable to asses that China has already developed a single season campaign capability against India on the continental borders. However, in the IOR, China remains highly vulnerable and would need decades of preparations for shaping the maritime space to undertake a full scale war against India.

Points to Ponder

A response to the increasing Chinese threat needs to put in place that not only will make the present but also the next generation of Chinese leadership weigh all pros and cons before any action but should preclude any military action all together, there by bringing peace between two emerging powers of Asia. Some responses purely from military point of view are as under.

- **Higher Defence Organization.** In the present form of the Higher Defence Organization in the country there is a total disconnect between the political class and the defence forces. This is further extenuated by a layer of bureaucracy in between. There is no representation of the defence forces in decision making at the national level. Comparing it with China or Pakistan, though both of them are totally different systems of governance, however when seen from the incorporation of defence in national decision making, certainly brings out some lessons for us to ponder over. Has India as nation lost out due to this deficiency? Is this system good and will deliver goods in a longer run? As democracy matures will this be a boon or bane? Is there a need to reorganize the MoD in a manner it responds to the needs of the defence forces as desired by the forces and the defence forces are part of decision making at the national level? An appraisal also need to be done by the defence forces if they are ready for this role and have a clear understanding of what is required of them? What kind of HDO we require once the triad is in place? As far as the Chinese threat is concerned, is there a need today to clearly evaluate the domain of decision; continental or maritime or both and accordingly work out our force development and modernization priorities?

- **Border Management.** The demarcation of disputed border is the responsibility of the Ministry of External Affairs, policing and over all border management is of MHA and in case of an external aggression the defence forces are responsible. Take the example of the LAC in Aksai Chin; as per various briefings of the MEA the LAC on ground is a perception, i.e., India and China does not agree on the same line on ground. The perceived LAC in behind each other's back – hence intrusions: flash points. To explain further – the Chinese project their area to be well into the area controlled by India and the vice versa is also true. Coming to the management of the LAC in Aksai Chin. At present it is being managed by both ITBP and the Army. The ITBP reports to the MHA and the Army reports to the MoD through the Army HQ. Before we talk of border settlement with the Chinese, won't it be prudent that we clearly demarcate the LAC before settling the larger border issue. The Kargil Review Committee and the GoM Report on National Security have clearly stated that there should be 'one border – one force' and till now we have not been able to implement this. In addition today is there need to integrate the 19 Command HQs into Integrated Commands responsible of geographical boundaries in both continental and maritime realms? Is it time now to rework the whole issue of Border Management including the command and control of border management forces? Is the cadre management of all the Border Management forces need a re look? Is the MHA and IPS the panacea to solve the present issues of border management?

- **Dual Task Formations (DTF).** The present policies of the DTFs need to be relooked both in terms of tasking and mobilization. The conflicts of future will not prolonged enough for Indian Army to effectively employ these formations if we keep the present parameters including mobilization as a yard stick. More over with continued RAPIDization of the formations along the Western border, if they were to be employed along the Northern borders, each division will be effectively a division minus. Also with the present system of infantry and mountain divisions, the inherent differences of weapon, equipment and vehicle holding profile ab-initio creates employment issues and combat readiness of the DTF formations. Is there a case in point to rework our system of DTFs? Do we need centrally placed strategic reserves with inherent lift capability? Is it time to re evaluate the threat from Pakistan and re assess the troops to task? Do we need to work out the requirement of mechanization?

- **Infrastructure Development.** The effort to develop infrastructure along the Western and Northern borders should be a continuous process and these projects should be national projects. The infrastructure must support the build up and maintenance of troops in both the phases i.e. border management and war. In addition it should also meet the aspirations of the local population and help in integrating into the national mainstream. Is infrastructure development along the borders considered a routine or it should be prioritized? If yes by whom? Is there a need to lay down all aspects of infrastructure development legislatively to include the lead agency to identify the need and planning, funding, clearances, execution and responsibilities?

- **Military Modernization.** It should be a continuous process based on the contours of threat perceptions those must be constantly monitored and regularly evaluated. The approach should be on capability and capacity building rather than a force of force approach. The defence forces of the nation should not only be capable of meeting all external and internal threats but also be equally capable of operating in an international environment with foreign militaries tackling both conventional and hybrid threats. However some important questions come to mind. Is the military modernization a slave of internal/ service and external interests? Why has the military modernization suffered in recent times? Is the finances and bureaucracy only responsible? Are the service HQs the "holy cows"? A detailed introspection is needed. Is indigenization the way forward? What timeframe, what fields, what level? How to weed out the deeply ingrained interests of the foreign suppliers?

Endnotes

1   Mao Zedong, "Problems of Strategy in China's Revolutionary War", Selected Works, Vol. I, pp. 190-91

2   The Economist, "The Dragon's New Teeth", April 7, 2012, accessed from http://www.economist.com/node/21552193, on March 23, 2014.

3   "The CPC had 85.13 million members at the end of 2012, according to a department statement on Sunday, one day before the 92nd anniversary of its founding", China's Communist Party membership exceeds 85 million, July 01, 2013, http://english.cpc.people.com.cn/206972/206974/8305636.html accessed on 24 May 2014

4   "Based on the total number of births, total number of deaths, net migration rates, and the population of 2013, the current population of the People's Republic of China is estimated to be about 1,390,510,630", Population of China 2014, February 19, 2014, http://www.worldpopulationstatistics.com/population-of-china-2014/ accessed on 17 May 2014

5   "Xi Jinping's Chinese Dream", Kuhn Robert Lawrence, The New York Times, June 4, 2013, http://www.nytimes.com/2013/06/05/opinion/global/xi-jinpings-chinese-dream.html?pagewanted=all&_r=1& accessed on 17 May 2014

6   India, China hold meet on border issue, The Hindu, February 10, 2014, New Delhi accessed from   http://www.thehindu.com/news/international/indiachinahold-meet-on-border-issue/article5673830.ece on July 24, 2014.

7   Ibid.

8   PTI, China calls Arunachal Pradesh 'Southern Tibet', Times of India, August 30, 2012, accessed from http://timesofindia.indiatimes.com/india/China-calls-Arunachal-Pradesh-southern-Tibet/articleshow/16005122.cms on July 3, 2013.

9   http://www.tehelka.com/why-india-would-lose-more-land-to-china/, September 06, 2013

10   ibid

11   Is China pushing Nepal to crack down on Tibetans? Sidner, Sara, February 22, 2012, http://edition.cnn.com/2012/02/21/world/asia/china-tibet-nepal/ accessed on March 24 2014

12   Settling differences with China, IDSA comment, Sanwal Mukul, May 7, 2014,   http://www.idsa.in/idsacomments/SettlingdifferenceswithChina_msanwal_070514 , accessed on May 24, 2014

13   ibid

14   Ibid

15   "Pakistan and China: Sweet as can be?" The Economist, May 12, 2011, http://www.economist.com/node/18682839 accessed on May 24, 2014

16   Annual Report to Congress, Military and Security Developments involving the Peoples Republic of China 2013, 77, Office of the Secretary of Defence, Department of Defence, USA

17   The Military Balance 2014, IISS, 236 – 239, Routledge, Feburary 2014

Conclusion

"Thus we may know that there are five essentials for victory:
He will win who knows when to fight and when not to fight.
He will win who knows how to handle both superior and inferior forces.
He will win whose army is animated by the same spirit throughout all its ranks.
He will win who, prepared himself, waits to take the enemy unprepared.
He will win who has military capacity and is not interfered with by the sovereign."

- Sun Tzu, The Art of War

"China does not pose a military threat. In fact, of all the major powers, China has probably been the most restrained in building up its military forces. China poses a very serious threat because it cannot be intimidated [by the US]."

- Noam Chomsky, 2007

Today China is sure that its time that she will achieve the glory and her place in the world which has been taken away from her by the Western powers. The so called 'Century of Humiliation' has ended and China is an established regional power and an emerging great power, both militarily and economically. Today China is the world's largest trading nation, with annual trade in goods worth $4 Trillion (tn) in 2013, second largest economy and the second largest defence spender with the third largest military. From a peasant's army in 1949, today PLA is the fastest modernizing military in the world, with no dearth of resources. How will the PLA look in 2025? Will it be a modern army with global reach or will it be contended to be a regional power with no or limited threat along her continental borders or her strategic rear and as far as the strategic front is concerned have an effective counter intervention mechanism and consolidate up to the first island chain for sure and attempt to reach up to the second island chain. Do the contours of her military modernization point towards a China that aspires to be a global power or is Chinese leadership still visualizes a China that will see world from a *tian xia* mind set or it sees a more inclusive world where China can cooperate with other major and rising powers of the region in mitigating and resolving existing and evolving security issues in the region and the world.

The present trends of PLA modernization indicate towards a China that will be a regional military power, with military capability, both organizationally and structurally, by 2025. It will certainly have a force projection and intervention capability in her immediate neighborhood primarily due to non availability adequate numbers of strategic air lifters and other force multipliers for force projection over long distances away from her borders and limited numbers of may be two or three, Carrier Battle Groups (CBGs). However the quality of the military will certainly be of a higher order than it is today.

Barring a major internal economic setback, it seems inevitable that Beijing will continue to increase the resources available to the military. But the force also faces constantly increasing costs for personnel, operations, and equipment. Unless official defence budgets are increased than those of the last decade, it is likely the PLA will continue to stress on economizing and the efficient use of centralized funds by its relatively large force, along with an unknown amount of support from local governments and an uncertain boost from other sources of income.

The PLA's missions may be modified as China grows and the international situation changes and require close attention by India and China's other neighbors. Fresh evidence must be gathered by close monitoring of trends and developments and regular analysis and assessments need to be undertaken to ensure that developments of the PLAA are understood in the context of overall PLAA modernization. Despite the paradigm changes that are underway in the other services, it appears the Army will likely continue to be the single largest major component of the PLAA, at least till 2025 and will remain a potent threat to its continental neighbors, especially India.

Appendix 1

Some identifications of formations and sub units in Chengdu and Lanzhou MAC[1]

Chengdu MAC	Lanzhou MAC
13th Group Army, Chongqing • 37th Division, Chongqing • 149th Light Mechanized Infantry Division, Leshan, Sichuan, RRF • UI Armoured Brigade, Pengzhou, Sichuan • UI Artillery Brigade, Chongzhou, Sichuan • UI AAA Brigade, Mianyang, Sichuan	**21st Group Army, Baoji, Shaanxi** • 61st Division("Red Army Division"), Tianshui, Gansu • *12th Armoured Division, Jiuquan, Gansu,RRF* • Artillery Brigade, Zhongnig, Ningxia • Air Defence Brigade, Linxia, Gansu
14th Group Army, Kunming, Yunnan • 31st Division, Dali, Yunnan, RRF • 40th Division, Kaiyuan, Yunnan, RRF • UI Armoured Brigade, Kunming, Yunnan • UI Artillery Brigade, Yunnan • UI AAA Brigade, Kunming, Yunnan • Chemical Defence Regiment	**47th Group Army, Lintong, Shaanxi** • 139th Mechanized Infantry Brigade, Weinan, Shaanxi • 55th Motorized Infantry Brigade, Zhangye, Gansu • 56th Motorized Infantry Brigade, Wuwei, Gansu • UI Armoured Brigade, Chengcheng, Shaanxi • Artillery Brigade, Gansu • AAA Brigade, Lintong, Shaanxi • Engineer Regiment • Communications Regiment

Units Subordinate to MAC or MD	Units Subordinate to MAC or MD
<ul><li>52nd Mountain Infantry Brigade, Nyingchi, Xizang, RRF</li><li>53rd Mountain Infantry Brigade, Nyingchi, Xizang</li><li>Army Aviation Regiment</li><li>"Cheetah" Special Operations Group, Chengdu, Sichuan</li><li>Special Reconnaissance Group, Chengdu, Sichuan (Technical Reconnaissance Bureau?)</li><li>Electronic Warfare Regiment</li><li>Technical Reconnaissance Bureau, Kunming, Yunnan</li><li>Chemical Defence Technical Group</li></ul>	<ul><li>4th Division ("Red Army Division"), Xinjiang MD, RRF</li><li>6th Mechanized Infantry Division, Hetian, Xinjiang MD, RRF</li><li>8th Motorized Infantry Division, Tacheng, Xinjiang MD</li><li>11th Motorized Division, Urumqi, Xinjiang MD</li><li>Artillery Brigade, Xinjiang MD</li><li>AAA Brigade, Xinjiang MD</li><li>Army Aviation Regiment, Xinjiang MD</li><li>Special Operations Dadui, Qingtongxia, Ningxia</li><li>ECM Regiment, Lanzhou, Gansu</li><li>High-Technology Reconnaissance Bureau, Lanzhou</li></ul>

Logistics Sub Departments and Units	Logistics Sub Departments and Units
• 22nd Sub department, Kunming, Yunnan • 37th Sub department, Chongqing • 38th Sub department, Chengdu, Sichuan	• 25th Sub department, Xining, Qinghai • 27th Sub department, Lanzhou, Gansu • 28th Sub department, Xi'an, Shaanxi • 29th Sub department, Xinjiang • 30th Sub department, Xinjiang • 31st Sub department • 32nd Sub department, Yecheng Xian, Xinjiang • Motor Transport Regiment, Xinjiang MD

Endnotes

1   Dennis J Blasko, "PLA Ground Force Modernization and Mission Diversification: Underway in all Military Regions "Right Sizing The People's Liberation Army: Exploring Contours of China's Military, pp 364-372.

Appendix 2

Deployment of PAPF in Chengdu and Lanzhou MAC[1]

Chengdu MAC	Lanzhou MAC
Border Defence	**Border Defence**
<ul><li>**Yunnan**</li><li>1st Border Defence Regiment, Funing.</li><li>2nd Border Defence Regiment, Malipo.</li><li>3rd Border Defence Regiment, Hekou.</li><li>4th Border Defence Regiment, Pingbian.</li><li>5th Border Defence Regiment, Jinping.</li><li>6th Border Defence Regiment.</li><li>7th Border Defence Regiment, Jiangcheng.</li><li>8th Border Defence Regiment, Mengla.</li><li>9th Border Defence Regiment, Diqing.</li><li>10th Border Defence Regiment, Lancang.</li><li>11th Border Defence Regiment, Cangyuan.</li><li>12th Border Defence Regiment, Luxi.</li></ul>	<ul><li>**Xinjiang**</li><li>1st Border Defence Regiment, Balikun.</li><li>2nd Border Defence Regiment, Qitai.</li><li>3rd Border Defence Regiment, Habahe.</li><li>4th Border Defence Regiment, Fuyun.</li><li>5th Border Defence Regiment, Emin.</li><li>6th Border Defence Regiment, Tuoli.</li><li>7th Border Defence Regiment, Bole.</li><li>8th Border Defence Regiment, Huochenghuiyuan.</li><li>9th Border Defence Regiment, Zhaosu.</li><li>10th Border Defence Regiment, Wushen.</li><li>11th Border Defence Regiment, Wuqia.</li><li>12th Border Defence Regiment, Tashenku'ergan.</li><li>13th Border Defence Regiment, Zepu.</li></ul>

	• Border Defence Battalion. • Kashen Border Defence Battalion, plus eight Border Defence Companies and one independent Border Defence Battalion.
Border Defence	
▪ **Xizang** • 1st Border Defence Regiment, Shannanlongzi. • 2nd Border Defence Regiment, Cuonei. • 3rd Border Defence Regiment, Dingri. • Border Defence Regiment, Changdu. • 5th Border Defence Regiment, Saga. • 6th Border Defence Regiment, Yadong. • Jiangzi Battalion. • Gangba 2nd Battalion. • Luozha 5th Battalion. • Milin Battalion. • Motuo Battalion.	
Internal Security	**Internal Security**
• Uncorroborated sources put the figure as 12 Mobile Divisions.	• Uncorroborated sources put the figure as 200,000 members of Xinjing Production and Construction Corps organized into 12 fully equipped paramilitary divisions.

Reserve formations/ units in the Chengdu and the Lanzhou MR[2]

Chengdu MR: Reserve Units	Lanzhou MR: Reserve Units
<ul><li>Sichuan Army Reserve Infantry Division, Chengdu</li><li>Sichuan Army Reserve AAA Division</li><li>Sichuan Leshan Reserve Infantry Brigade</li><li>Sichuan Dazhou Army Reserve Artillery Brigade</li><li>Chongqing Reserve AAA Division</li><li>Chongqing Army Reserve Logistics Support Brigade</li><li>Guizhou Army Reserve Infantry Division</li><li>Yunnan Army Reserve Infantry Division</li><li>Tibet Army Reserve Mixed Brigade</li></ul>	<ul><li>Shaanxi Army Reserve 141st Infantry Division</li><li>Shaanxi Army Reserve AAA Division, Xi'an</li><li>Hanzhong Reserve AAA Regiment (possibly subordinate to the Shaanxi AAA Division)</li><li>Shaanxi Baoji Army Reserve Logistics Support Brigade</li><li>Gansu Army Reserve Infantry Tianshui Brigade</li><li>Gansu Army Reserve AAA Division, Lanzhou</li><li>Xinjiang Army Reserve Infantry Division</li><li>Urumqi Reserve AAA Regiment (possibly subordinate to the Xinjiang Army Reserve Infantry Division)</li><li>Qinghai Army Reserve Infantry Division</li><li>Qinghai Army Reserve Infantry Brigade</li><li>Yinchuan Reserve AAA Regiment</li></ul>

Endnotes

1 Dennis J Blasko, "PLA Ground Force Modernization and Mission Diversification: Underway in all Military Regions "Right Sizing The People's Liberation Army: Exploring Contours of China's Military, pp 364-372.

2 Ibid.

Appendix 3

Present Deployment of the PLAAF[1]

At present the PLAAF, as part of the seven Military Area Commands (MACs), is organized in seven Military Region Air force (MRAF) for command and control. Subordinate to the MRAF are the five Corps Command Posts (Deputy Leader grade), which are Datong, Kunming, Wuhan, Xian and Fuzhou, four Corps bases (Deputy Leader grade), which are Nanning, Urumqi, Shanghai and Dalian and four Division Leader grade Command Posts at Lhasa, Hetian, Zhangzhou and Changchun. All of these are responsible for all subordinate combat organizations in its area of operations.

Shenyang MRAF

1ˢᵗ Fighter Division	
Ftr Regt with J-11B	1
FGA Regt with J-10A	1
Ftr Regt with J-8F	1
11ᵗʰ Attack Division	
Atk Regt with JH-7A	1
Atk Regt with Q-5	1
16ᵗʰ Special Mission Division	
EW Regt with Y-8cB/G	1
ISR Regt with JZ-8F	1
Tpt Regt with Y-8	1
21ˢᵗ Fighter Division	
Ftr Regt with J-7E	1
Ftr Regt with J-8H	1
Ftr Regt with J-7H	1
Dalian Base	
FGA Bde with J-7E; J-11B; Q-5	2

Harbin Flying Academy	
Trg Bde with CJ-6; JL-8; Y-7	2
Other Forces	
(mixed) SAM/ADA Bde	1
Sam Bde	1

North Sea Fleet Naval Aviation

Trg Regt with CJ-6A	1
Trg Regt with HY-7	1
Trg Regt with Y-5	1

Beijing MRAF

7th Fighter Regt	
Ftr Regt with J-11	1
Ftr Regt with J-7G	1
Ftr Regt with J-7	1
15th Fighter/Attack Division	
FGA Regt with J-10A	1
Atk Regt with Q-5C	1
24th Fighter Division	
Ftr Regt with J-8F	1
FGA Regt with J-10A	1
Shijizhuang Flying Academy	
Trg Bde with CJ-6; JL-8; Y-5; Y-7	1
Flight Test Centre	
with J-7/J-8/J-10/JL-9/Su-30 (on rotation)	1
SAM Div	3
(mixed) SAM/ADA Div	1

Other Forces

34ᵗʰ VIP Transport Division	
Tpt Regt with B-737; CRJ 200/700	1
Tpt Regt With B-737; Tu-154M; Tu-154M/D	1
Tpt Regt with Y-7	1
Hel Regt with AS332	1

North Sea Fleet Naval Aviation

7ᵗʰ Naval Air Division	
Trg regt with JL-9	1
FGA Regt with JH-7A	1
Tpt Regt with Y-7/Y-8	1
Rg Regt with HY-7	1
Trg Regt with Jl-8	1

Lanzhou MRAF

6ᵗʰ Fighter Division	
Ftr Regt with J-11	1
Ftr Regt with J-7E	1
Ftr Regt with J-7	1
36ᵗʰ Bomber Division	
Surv Regt with J-7E	1
Bbr Regt with H-6M	1
Bbr Regt with H-6H	1
Urumqi Base	
FGA Bde with J8-H; J-11B; JH-7A	2
Other Forces	
Trg Regt with J-7/JJ-7	1
Trg Regt with JL-8	4
Trg Regt with Y-7	1

Hel Trg Regt with Z-9	1
(mixed) SAM/ADA Div	1
Indep SAM Regt	4

Jinan MRAF

5th Attack Division	
Atk Regt with Q-5E	1
Atk Regt with JH-7A	1
12th Fighter Division	
Ftr Regt with J-8B	1
Ftr Regt J-7G	
19th Fighter Division	
Ftr Regt with Su-27SK	1
Ftr Regt with J-7	1
Trg Regt with J-7/JJ-7	1
Other Forces	
Flight Instructor Training Sane with CJ-6; JL;8	1
SAM Bn	4

North Sea Naval Aviation

5th Naval Air Division	
FGA Regt with JH-7A	1
Ftr Regt with J-8F	1
Other Forces	
EW/ISR/AEW Regt with Y-8J/JB/W	1
MP Regt with SH-5	1
Hel Regt with AS365; Ka-28; SA 321; Z-9	1

Nanjing MARAF

3rd Fighter Division	
Ftr Regt with H-6H	1
FGA Regt with J-10A	1
FGA Regt with Su-30MKK	1
10th Bomber Division	
Bbr Regt with H-6H	2
EW Regt with Y-8CB/G/XZ	1
14th Fighter Division	
Ftr Regt with J-11	1
Ftr Regt with J-7E	1
26th Special Mission Division	
AEW&C Regt with KJ-200/KJ2000/Y-8T	1
CSAR Regt with M-171/Z-8	1
28th Attack Division	
Atk Regt with JH-7A	2
Atk Regt with Q-5D/E	1
32nd Fighter Division	
Ftr Regt with JH-7A	1
Trg Regt with J-7/JJ-7	1
Shanghai Base	
FGA/ISR Bde with J-7E; J-8H; JZ-8F; Su-30MKK	2
Other Forces	
Sam Bde	3
ADA Bde	1
Indep SAM Regt	2

East Sea Fleet Naval Aviation

4th Naval Aviation Division	
FGA Regt with Su-30 Mk2	1
FGA Regt with J-10A	1

6th Naval Aviation Division	
FGA Reg with JH-7	2
bbr Regt with H-6G	1
Other Forces	
Hel Regt with Mi-8; Ka-28; Ka-31	1

Guangzhou MRAF

2nd Fighter Division	
Ftr Regt with J-8H	1
FGA Regt with J-10A	1
Ft Regt with J-11	1
8th Bomber Division	
Tkr Regt with H-6U	1
bbr Regt with H-6H	1
bbr Regt with H-6K	1
9th Fighter Division	
FGA Regt with J-10A	1
Ftr Regt with J-7E	2
13th Transport Division	
Tpt Regt with Y08	1
Tpt Regt with Y07	1
Tpt Regt with II-76MD/TD	1
18th Fighter Division	
Ftr Regt with J-7	1
FGA Regt with Su-30MKK	1

Nanning Base	
Ftr Bde with J-7/J-7G Fishbed	2
Other Forces	
SAM Bde	4
ADA Bde	1
Indep ADA Regt	1
Other Forces	
Marines	
Sam Bde	2
15th Airborne Corps	
AB Div	3

South Sea Fleet Naval Aviation

8th Naval Aviation Division	
FGA Regt with J-11B	1
bbr Regt with H-6G	1
Ftr Regt with J-7E	1
9th Naval Aviation Division	
Ftr Regt with J-8H	1
FGA Regt with JH-7A	1
Other Forces	
Tpt Regt with Y-7; Y-8; Z-8; Z-8JH/S; Z-9	1

Chengdu MRAF

4th Transport Division	
Tpt Regt with Y-8/Y-9	1
Tpt Regt with Y-7	1
Tpt Regt with Mi-17V-5/Y-7	1
33rd Fighter Division	
Ftr Regt with J-7E	1
Ftr Regt with J-11	1

44th Fighter Division	
Ftr Regt with J-7	1
FGA Regt with J-10	1
Other Forces	
(mixed) SAM/ADA Bde	1
Indep SAM Regt	1

Endnotes

1   IISS, The Military Balance 2013

Appendix 4

Modernization Trends PLAAF 2009 Onwards[1]

2009		2011		2013		Remarks
Shenyang MRAF						
1st Fighter Division(Anshan)		1st Fighter Division		1st Fighter Division		
1st Regt (Anshan)	J-11/ J-11B	1 Ftr Regt	J-11/ J-11B	1 Ftr Regt	J-11B	
2nd Regt (Anshan)	J-8F	1 FGA Regt	J-10A	1 FGA Regt	J-10A	
3rd Regt (Anshan)	J-8A	1 Ftr Regt	J-8F	1 Ftr Regt	J-8F	
11th Atk Division (Jilin, Siping)		11th Atk Division		11th Attack Division		
31st Regt	Q-5D	1 ATk Regt	JH-7A	1 Atk Regt	JH-7A	
33rd Regt	Q-5D	1 ATk Regt	Q-5D	1 Atk Regt	Q-5	
				16th Special Mission Division		
				1 EW Regt	Y-8 CB/G	
				1 ISR Regt	JZ-8F	
				1 Tpt Regt	Y-8	
21st Fighter Division		21st Fighter Division		21st Fighter Division		
61st Regt (Heilong jiang, Mudan jiang)	J-8E	1 Ftr Regt	J-7E	1 Ftr Regt	J-7E	
62nd Regt (Qiqihar)	J-7	1 Ftr Regt	J-8H	1 Ftr Regt	J-8H	
63rd Regt (Mudan jiang)	J-8B	1 Ftr Regt	J-8B	1 Ftr Regt	J-7H	

30th Fighter Division(incl civil & PLAAF base-Liaoning, Dalian)		30th Fighter Division			
88th Regt	J-8A/E	2 Ftr Regt	J-7E		
89th Regt	J-8F	1 Ftr Regt	J-8E		
90th Regt	J-7E				
4th Recce Regt with JZ-8		1 Indep Recce Regt with JZ-8			
				Dalian Base: 2 FGA Bde's with J-7E, J-11B & Q-5	
MRAF Training Bases • 1st Air Academy: Heilong jiang, Harbin with • 1st Ai		3 Training Schools with An-30/ CJ-6/ H-5/ HJ-5/ JJ-5/ K-8/ Y-7		Harbin Flying Academy: 2 Trg Bde's with CJ-7, JL-8, Y-5 & Y-7	
		Other Forces: 1 (mixed) AD bde & SAM bde		Other Forces: 1 (mixed) SAM/ ADA bde & SAM bde	
Beijing MRAF					
7th Fighter Division		7th Fighter Division		7th Fighter Division	
19th Regt (Zhangjiakou)	J-11	1 Ftr Regt	J-11	1 Ftr Regt	J-11B
21st Regt (Zhangjiakou)	J-7B	1 Ftr Regt	J-7G	1 FGA Regt	J-7G
				1 Ftr Regt	J-7
15th Fighter/ Attack Division		15th Fighter/ Attack Division		15th Fighter/ Attack Division	
43rd Regt (Hairen)	J-7C	1 Atk Regt	Q-5C	1 Atk Regt	Q-5C
44th Regt (Lingqiu)	Q-5	2 Ftr Regt	J-7	1 FGA Regt	J-10A
45th Regt	J-7D				
24th Fighter Division		24th Fighter Division		24th Fighter Division	

70th Regt (Yangkun, Tianjin)	J-8 A/E	1 Ftr Regt	J-8	1 Ftr Regt	Q-5C	
71st Regt (Zunhua)	J-8	1 Ftr Regt	J-10A	1 Ftr Regt	J-10A	
72nd Regt (Yangkun)	J-8					
				Shijiazhuang Flying Academy: 1 Trg Bde with CJ-6, JL-8, Y-5 & Y-7		
		1 Trg Base	J-7B; JJ-7			
		2 Trg Schools	CJ-6/ JJ-5/ K-8			
Other Forces		Other Forces		Other Forces		
Flight Test & Training Centre (Hebei, Cangzhou)	4 Regts with Su-30/ Su-27/ J-11/ J-8C/ J-10/ J-7E/JJ-7 (on rotation)	1 Flight Test Centre	Su-30/ Su-27/ J-11/ J-8C/ J-10/ J-7E/JJ-7 (on rotation)	1 Flight Test Centre	J-7/ J-8/ J-10/ JL-9/ Su-30 (on rotation)	
MRAF Training Base; 1st Air Academy (Hebei, Tangshan)	J-7B; JJ-7					
MRAF Training Base; 4th Air Academy (Hebei, Shijiazhuang/ Baoding)	4 Regts with K-8, JJ-5 & CJ-6					
MRAF Training Base; 6th Air Academy (Hebei, Zhuzhou)	3 Regts with JJ-5 & CJ-6					

Air Defence		Air Defence		Air Defence		
3 SAM Div	• 5th Beijing with S-300, HQ-9 & HQ-2	3 SAM Div & 1 (mixed) SAM/ ADA Div		3 SAM Div & 1 (mixed) SAM/ ADA Div		
	• 6th Beijing HQ-7 & HQ-2					
1 SAM Regt	• 7th Tian-jin with HQ-2 & AAA					
Indep AAA Regt at Guangdong	• 7th Guang xi, Liaon-ing with HQ-2					
				34th VIP Transport Division		
				1 Tpt Regt	B-737; CRJ 200/ 700	
				1 Tpt Regt	B-737; Tu-154M; Tu-154M/ D	
				1 Tpt Regt	Y-7 & Hel Regt AS 332	

Lanzhou MRAF

6th Fighter Division		6th Fighter Division		6th Fighter Division		
16th Regt (Ningxia, Yinchuan)	J-11	1 Regt	J-11	1 Regt	J-11	
18th Regt (Yinchuan)	J-7E	1 Regt	J-7E	1 Regt	J-7E	

139th Regt (Tianshui)	J-7	1 Regt	J-7	1 Regt	J-7	
36th Bomber (bbr) Division (Shangxi, Wugong/Lintong)		36th Bomber (bbr) Division		36th Bomber (bbr) Division		
106th Recce/ Bbr Regt (Hairen)	Y-8H-1 Aerial Surv Regt	1 Surv Regt	Y-8H	1 Surv Regt	Y-8H-1	
107th Regt (Lingqiu)	H-6E	1 Bbr Regt	H-6M	1 Bbr Regt	H-6M	
108th Regt	H-6A	1Bbr Regt	H-6H	1Bbr Regt	H-6H	
37th Fighter Division (Xinjiang, Urumqi)		37th Fighter Division				
109th Regt	J-8H	1 Ftr Regt	J-8H			
110th Regt	J-7E	1 Ftr Regt	J-7E			
111st Regt	J-7G	1 Ftr Regt	J-7G			
				Urumqi Base; 2 FGA Bdes	J-8H; J-11B; JH-7A	
MRAF Air Academy						
5th Air Academy (Gangsu, Wuwei)	3 Regts with JJ-5 & CJ-6	2 Trg Schools	CJ-6 & JJ-5	1 Trg Regt	J-7/ JJ-7	
8th Air Academy (Xinjian, Shuquan)	3 Regts with JJ-5 & CJ-6			4 Trg Regts	JL-8	
PLAAF Missile Testing Regiment (Gansu, Dinxi)	JJ-6 & J-7B	PLAAF Missile Testing Regiment	JJ-6 & J-7B	1 Trg Regt	Y-8	

Air Defence		Air Defence		Air Defence		
6th SAM Bde (Gansu, Lanzhou)	HQ-2 & AAA	1 Mixed AD Division	NK	1 Mixed SAM/ ADA Division	NK	
6th SAM Regt (Xinjian)	HQ-2E	1 SAM Bde	NK	1 SAM Bde	NK	
41st Radar Regt (Xinjian, Hetian)			NK		NK	
		4 Indep SAM Regts		4 Indep SAM Regts		

Jinan MRAF

5th Atk Division (Shangdong)		5th Atk Division		5th Attack Division		
13th Regt	Q-5E	1 Atk Regt	Q-5E	1 Atk Regt	Q-5E	
14th Regt	JH-7A	1 Atk Regt	JH-7A	1 Atk Regt	JH-7A	
12th Fighter Division (Shangdong, Jinan)		12th Fighter (Ftr) Division		12th Fighter (Ftr) Division		
34th Regt	J-8B	2 Ftr Regt	J-8B	1 Ftr Regt	J-8B	
35th Regt	J-8B	1 Ftr Regt	J-7G	1 Ftr Regt	J-7G	
36th Regt	J-8B					
19th Fighter Division (Henan, Zhengzhou)		19th Fighter (Ftr) Division		19th Fighter (Ftr) Division		
55th Regt	J-11	1 Ftr Regt	Su-27SK	1 Ftr Regt	Su-27SK	
56th Regt	J-11	1 Ftr Regt	J-7	1 Ftr Regt	J-7	
57th Regt	J-7E	1 Ftr Regt	J-7E	1 Trg Regt	J-7/ JJ-7	
1st Recce Regt (Shangdong, Wenteng)	JZ-6	1 Indep Recce Regt	JZ-6			

				1 Flight Instructor Trg Base	CJ-6 & JL-8	
		Air Defence	4 SAM Bn	Air Defence	4 SAM Bn	
Nanjing MRAF						
3rd Fighter Division		3rd Fighter (Ftr) Division		3rd Fighter (Ftr) Division		
7th Regt	J-10	1 Fighter Ground Attack Regt	J-10	1 Fighter Ground Attack Regt	J-10A	
8th Regt	J-7E	1 Ftr Regt	J-7E	1 Ftr Regt	J-7G	
9th Regt	Su-30MKK	1 Fighter Ground Attack Regt	Su-30 MKK	1 Fighter Ground Attack Regt	Su-30MKK	
10th Bomber (bbr) Division		10th Bomber (bbr) Division		10th Bomber (bbr) Division		
28th Regt (Anqing)	H-6E	2 Bbr Regts	H-6E	2 Bbr Regts	H-6H	
29th Regt (Nanjing)	H-6E	1 EW Regt	Y-8D	1 EW Regt	Y-8 CB/ G/XZ	
30th Regt (Nanjing)	Y-8D					
14th Fighter Division (Jiangxi, Nangchang, Xiangtang)		14th Fighter (Ftr) Division		14th Fighter (Ftr) Division		
40th Regt (Nangchang, Xiangtang)	J-11A	1 Ftr Regt	J-11	1 Ftr Regt	J-11	
41st Regt (Jiojiang)	J-7H	1 Ftr Regt	J-7E	1 Ftr Regt	J-7E	
26th Special Mission Division (Wuxi, Jiangsu, Suzhou, Guangfu)		26th Special Mission Division		26th Special Mission Division		

1 Regt	KJ-2000	1 AEW & C Regt	KJ-2000/ KJ-200	1 AEW&C Regt	KJ-2000/ KJ-200/ Y-8T
1 Regt (recce)	JZ-8F	1 Recce Regt	JZ-8F		
		1 Combat Search & Rescue (CSAR) Regt	M- 171 / Z -8	1 Combat Search & Rescue (CSAR) Regt	M-171/ Z-8
28th Atk Division (Zhejiang, Hangzhou)		28th Atk Division		28th Atk Division	
82nd Regt	JH-7A	1 Atk Regt	JH-7A	2 Atk Regt	JH-7A
83rd Regt	Q-5D	2 Atk Regts	Q-5D	1 Atk Regts	Q-5D
84th Regt	Q-5D				
29th Fighter Division (Zhejiang, Quzhou)		29th Fighter (Ftr) Division			
85th Regt	Su-30MKK	1 FGA Regt	Su-30 MKK		
86th Regt	J-7C	1 Ftr Regt	J-11		
87th Regt	J-8B/D	1 Ftr Regt	J-8D		
				32nd Fighter (Ftr) Division	
				1 Ftr Regt	J-11B
				1 Trg Regt	J-7/ JJ-7
				Shanghai Base	
				2 FGA/ ISR Bde	J-7E; J-8H; JZ-8F; Su-30MKK

MRAF Trg Base						
3rd Recce Regt	JZ-6					
13th Air Academy (Anhui, Bingbu)	K-8, JJ-5 & CJ-6	1 Trg School	K-8, JJ-5 & CJ-6			
Air Defence		Air Defence		Air Defence		
2nd SAM Bde (Fujian, Xianyou)	S-300 & HQ-9	3 SAM Bdes		3 SAM Bdes		
3rd SAM Bde (Shanghai)	HQ-7 & HQ-2	1 ADA Bde		1 ADA Bde		
4th SAM Bde (Jiangsu, Nanjing)	HQ-9 & HQ-7					
8th SAM Bde (Shanghai)		2 Indep SAM Regts		2 Indep SAM Regts		
	HQ-9					
Guangzhou MRAF						
2nd Fighter Division (Guangdong, Zhanjiang Suxi)		2nd Fighter (Ftr) Division		2nd Fighter (Ftr) Division		
4th Regt (Liuzhou)	J-7B	1 Ftr Regt	J-7G	1 Ftr Regt	J-8H	
5th Regt (Qifengling)	J-10	1 FGA Regt	J-10	1 FGA Regt	J-10A	
6th Regt (Zhangjiang Suxi)	J-11	1 Ftr Regt	Su-27SK/ J-11	1 Ftr Regt	J-11	
8th Bomber (bbr)/ Tanker (tkr) Division (Hunan, Leiyang)		8th Bomber (bbr)/ Tanker (tkr) Division		8th Bomber (bbr)/ Tanker (tkr) Division		
22nd Regt (Liuzhou)	H-6E	1 Tkr Regt	H-6U	1 Tkr Regt	H-6U	
23rd Regt (Liuzhou)	H-6U	1 Bbr Regt	H-6H	1 Bbr Regt	H-6H	
		1 Bbr Regt	H-6E	1 Bbr Regt	H-6K	
9th Fighter Division (Guangdong, Shaoguan)		9th Fighter (Ftr) Division		9th Fighter (Ftr) Division		

25th Regt	J-8B	1 FGA Regt	J-10	1 FGA Regt	J-10A
26th Regt	J-8D	2 Ftr Regt	J-8D	2 Ftr Regt	J-7E
27th Regt	J-7B				
13th Airlift Division (Wuhan, Wangjiadun)		13th Airlift Division		13th Transport Division	
37th Regt (Henen, Kaifeng)	IL -76 MD	2 Transport (Tpt) Regts	IL-76 MD	1 Tpt Regt	IL-76 MD/TD
38th Regt (Wuhan, Wangjiadun)	An-26/ Y-8	1 Tpt Regt	Y-7/ Y-8	1 Tpt Regt	Y-8
39th Regt (Hubei, Dunyang)	IL -76 MD			1 Tpt Regt	Y-7
18th Fighter Division (Hunan, Changsha)		18th Fighter (Ftr) Division		18th Fighter (Ftr) Division	
52nd Regt (Wuhan, Shangpo)	J-7B	1 Ftr Regt	J-7	1 Ftr Regt	J-7
54th Regt (Changsha, Datuopu)	Su-30 MKK	1 FGA Regt	Su-30 MKK	1 FGA Regt	Su-30 MKK
				Nanning Base	
				2 Ftr Bdes	J-7/ J-7G *Fishbed*
42nd Fighter Division (Guangxi, Nanning)		42nd Fighter (Ftr) Division			
124th Regt (Baicetianyang)	J-7B	2 Ftr Regts	J-7		
125th Regt (Nanning)	J-7H				
2nd Indep Recce Regt (Jiangxi, Taihe)	JZ-6	1 Indep Recce Regt	JZ-6		

Air Defence		Air Defence		Air Defence		
9th Sam Bde (Hunan, Huaibei)	HQ-2 & AAA	4 SAM Bdes		4 SAM Bdes		
10th Sam Bde (Guangdong, Guangzhou)	HQ-2	1 ADA Bde		1 ADA Bde 1 Indep ADA Regt		
11th Sam Bde (Guangdong)	HQ-7 & AAA	1 Indep ADA Regt				
7th Sam Bde (Guangxi, Nanning)	HQ-7 & AAA					
Indep AAA Regt (Guangdong, Huangpu)	AAA					
				Other Forces		
				15th Air Borne Corps	3 AB Divs	

Chengdu MRAF

4th Airlift Division (Chengdu, Qionglai)	An-26/ Mi-17 Helicopter	4th Airlift Division	Y-7/ Mi-17	4th Airlift Division		
				1 Tpt Regt	Y-8/ Y-9	
				1 Tpt Regt	Y-7	
				1 Tpt Regt	Mi-17V5/ Y-7	
33rd Fighter Division (Chongqing, Baishiyi)		33rd Fighter (Ftr) Division		33rd Fighter (Ftr) Division		
97th Regt	J-7E	1 Ftr Regt	J-7E	1 Ftr Regt	J-7E	
98th Regt	J-7B	1 Ftr Regt	J-7	1 Ftr Regt	J-11	

99th Regt	J-11	1 Ftr Regt	J-11			
44th Fighter Division (Yunnan, Kunming)		44th Fighter (Ftr) Division		44th Fighter (Ftr) Division		
130 th Regt	J-7B	2 Ftr Regts	J-7	1 Ftr Regt	J-7	
131th Regt	J-7B	1 FGA Regt	J-10	1 FGA Regt	J-10	
132nd Regt	J-10					
2nd Air Academy (Sichuan, Jiajiang)	H-5, HJ-5 & CJ-6	1 Trg School	H-5, HJ-5 & CJ-6			
		Air Defence		Air Defence		
		1 (Mixed) AD Bde		1 (Mixed) SAM/ ADA Bde		
		3 Indep SAM Regts		3 Indep SAM Regts		

Endnotes

1   The data has been extracted from The Military Balance published by IISS in 2009, 2011 and 2013.

Appendix 5

Transformation of PLAN from 2007 to 2013

		2007	2009	2011	2013
Submarines		**58**	65		65
Strategic					
SSBN	• Xia Class	**1**	1	1	1
	• Jin Class		2	2	3
Tactical		**57**	**62**	68	61
SSN	• Han Class (Type 091)	4	4	4	3
	• Shang Class (Type 093)		2	2	2
SSG	Modified Romeo (S5G) Class	1	1	1	
SSK		50	54	60	55
	• Kilo Class	10	12	12	12
	• Ming Class (Type ES5C/D)	1			
	• Ming Class	19	19	20	20
	• Improved Ming Class (Type ES5E)	19	19		
	• Type-035			4	4
	• Type-035G			12	12
	• Type-035B			4	4
	• Romeo Class (Type ES3B/ Type-033)	15	8	8	
	• Song Class (Type-039/ 39G)	9	13	16	16
	• Yuan Class (Type-39 A/B)	2	2	4	4
	• Yuan Class II (Type 39 B)				3
SS	Golf Class (SLBM Trials)	1	1	1	
SSB	Qing Class (SLBM Trials)				1

Principal Surface Combatants		57	78	78	
Aircraft Carrier (CV)					1
Destroyers		28	28	13	14
	• Guangzhou Class	2	-		
DDGHM	• Hangzhou (RF Sovremeny) • Class	3	4	4	4
	• Luyang Class	-	2	2	2
	• Luyang II	-	2	2	3
	• Lanzhou Class	2	2		
	• Luda III Class	1	1		
	• Luda (Type 051) Class	11	10		
	• Luhai Class	1	1	1	1
	• Luhu (Type 052A/52) Class	2	2	2	2
DDGM	• Luzhou (Type 051C) Class (DDGM)	1	-	2	2
	• Luda II	1	1		
	• Luda Mod Type 051DT	4	3		
Frigates		48	50	65	62
FFGHM					
	• Jiangkai (Type 054) Class	2	2	2	2
	• Jiangkai II Type 054A) Class		4	7	13
	• Jiangwei (Type 053 H2G) Class	4	4	4	4
	• Jiangwei II (Type 053H3) Class	10	10	10	10
FFGH	• Jianghu IV(Type 053H1Q) Class	1	1	1	1
FFGM	• Luda III (Type 051DT) Class			2	2
	• Luda III (Type 051G)				2
FFG					28

	• Jianghu I (Type 053H) Class	13	11	11	8
	• Jianghu II (Type 053H1) Class	9	9	8	5
	• Jianghu III (Type 053H2) Class	3	3	3	3
	• Jianghu V (Type 053H1G) Class	6	6	6	6
	• Luda (Type 051/ 051D/ 051Z) Class			9	6
	• Luda II (Type 051G) Class			1	
	• Luda III (Type 051G II) Class			1	
Patrol & Coastal Combatants		**242**	**247**		**211+**
Fast Patrol Craft with SSM (PFM)/ Patrol Craft with Guided Missile (PCG)		42	77	26	
	• Houxin Class (Type 037/ IG)	16	16	20	
	• Houjian Class (Type 037/ II)			6	
	• Huang Class	7	7		
Patrol Craft Fast with Guided Missile (PCFG)				76	76
	• Huangfeng/ Hola • (FSU Osa- IType) Class (Type 021)	15	14	11	11
	• Houbei Class (Type 022)	4	40	65	65+
Fast Patrol Craft Coastal (PFC)	Hainan Class	98	93		
Patrol Craft Coastal (PCC)		22	27	75	75
	• Haijui Class (Type 037/I)	2	2	3	3
	• Hainan Class (Type 037)			50	50

	• Haiqing Class (Type 037/IS)	20	25	22	22
Patrol Craft Inshore (PCI)/ Patrol Boat (PB)		50	50	34	34
	• Haizui Class (< 100 T)	15	15	34	34
	• Shanghi Class (< 100T)	35	35		
Patrol Craft Riverine (PCR) less than 100 T		30	-		
Mine Warfare		**65**	69	73	47
Mine Counter Measures		64	68	88	46
• Mine Counter Measure Vessels (MCMV)/ Mine Counter Measure Ocean (MCO)			4	7	
	• Wochi Class		4	6	6
	• Wozang Class			1	1
• Mine Sweeper Coastal (MSC)	Wosao Class	4	4	16	**16**
• Mine Sweeper Drone Inshore (MSD)	• Futi Class (Type 312)	4	4	4	4
	• Reserve	42	42	42	42
	• Others	-	-	3	3
• Mine Sweeper Ocean (MSO)	T-43	14	14	16	16
• Mine Layer (ML)		1	1	1	1
Amphibious					
Landing Port Dock	Yuzhao Class (Type 071) (2 LCAC/ 4 UCAC or ACV, 500-800 troops & 2 helicopters)		1	1	2
Landing Ship (LS)		73	83	87	85

• Landing Ship Medium (LSM)		47	56	61	59
	• Yubei Class (10 tanks or 150 troops)			10	10
	• Yudao Class	1	1	-	
	• Yudeng Class (6 Tanks; 180 troops)	1	1	1	1
	• Yuhai Class (2 Tanks; 250 troops)	13	13	10	10
	• Yuliang Class (5 Tanks; 250 troops)	22	31	30	28
	• Yunshu Class (6 Tanks)	10	10	10	10
• Landing Ship Tank (LST)		26	27	26	26
	• Yukan Class (10 Tanks; 200 troops)	7	7	7	7
	• Yuting Class (10 Tanks; 250 troops; helicopter landing platform)	10	10	9	9
	• Yuting II Class (4 LCVP; 10 Tanks; 250 troops)	9	10	10	10
Landing Craft (LC)		160	160	151	151
• Landing Craft Utility (LCU)		130	130		120
	• Yubei Class (10 Tanks or 250 troops)	10	10		
	• Yunnan Class	120	120	120	120
• Landing Craft Medium (LCM)	• Yuchin Class	20	20	20	20
• Air Cushion Vehicle (ACV)	• UCAC	10	10	10	10
	• LCAC			1	1
Logistics and Support		163	204	205	205

• Logistics Ship (AFS)		14			
• Miscellaneous Auxiliary (AG)/ Hospital Ship (AH)		6	6	7	6
	• Qiongsha Class (capacity 400 troops) (AG)		4	4	4
	• Qiongsha Class (hospital conversion) (AG)		2	2	2
	• Daishan Class (AH)			1	1
• Tanker with Helicopter Capacity (AORH)			5	5	5
	• Fuqing Class		2	2	2
	• Fuchi Class		2	2	2
	• Nanyun Class		1	1	1
• Ice Breaker (AGB)		4	4	4	4
	• Yanbing Class	-	1	1	1
	• Yanha Class	-	3	3	3
• Oceanographic Research Vessel (AGOR)		33	5	5	5
	• Dahua Class		1	1	1
	• Kan Class		2	2	2
	• Bin Hai Class		1	1	1
	• Shuguang Class		1	1	1
• Intelligence Collection Vessel (AGI)	Dadei Class		1	1	1
• Space & Missile Tracking Ship (AGM)		-	5	5	5

• Sea-going Buoy Tender Ship (ABU)	Yannan Class	-	7	7	7
• Survey Ship (AGS)			6	6	6
	• Yenlai Class		5		5
	• Ganzhu Class		1		1
• Tanker with RAS Capability (AOL)		3	5		5
	• Fuqing Class	2	-		
	• Nanchang Class	1	-		
	• Guangzhou Class		5	5	5
• Tanker (AOT)		33	50	50	50
	• Danlin Class		7	7	7
	• Fulin Class		20	20	20
	• Shengli Class		2	2	2
	• Jinyou Class		3	3	3
	• Fuzhou Class		18	18	18
• Water Tanker (AWT)		-	-	18	18
	• Leizhou Class			10	10
	• Fuzhou Class			8	8
• Repair Ship (AR/ AS)		12	8	8	8
	• Dazhi Class		1	1	1
	• Dalang Class		5	5	5
	• Dazhou Class		2	2	2
• Submarine Rescue Craft (ASR)		1	1	1	1
	• Daijiang Class with two SA-321 Super Frelon helicopters	-	1	1	1

• Rescue & Salvage Ship (ARS)				2	2
	• Dadong Class			1	1
	• Dadao Class			1	1
• Tug, Ocean Going (ATF)		25	51	51	51
	• Tuzong Class		4	4	4
	• Hujiu Class		10	10	10
	• Daozha Class		1	1	1
	• Gromovoy Class		17	17	17
	• Roslavl Class		19	19	19
• Cargo Ship (AK)		23	23	23	23
	• Yantai Class		2	2	2
	• Dayun Class		2	2	2
	• Danlin Class		6	6	6
	• Dandao Class		7	7	7
	• Hongqi		6	6	6
• Transport Ship (Tpt)		30			30
• Training Ship (One Helicopter training) (TRG)		2	2	2	
	• Shichang Class		1	1	
	• Daxin		1	1	
• Degaussing Ship (YDG)	Yen Pai Class		5	5	5
Naval Aviation		26,000	26,000	26,000	26,000
Aircraft; Combat Capable		792	290	311	341
• Bomber (BBR)		130	50	50	30
	• H-5, F-5, F-5B (Il-28) Beagle (torpedo carrying light bomber)	100	20		
	• H-5			20	

	• H-6D	30	30		
	• H-6G			30	30
• Fighter (FTR)		346	84	84	72
	• J-8I/ J-8F/J-8B/J-8D Finback		48		
	• J-8F/ J-8H			48	24+24
	• J-8B Finback	50	-		
	• J-8D Finback	20	-		
	• J-8/ J-8A Finback	50	-		
	• J-8 II A Finback	200	-		
	• J-7/ J-7E Fishbed			36	24
	• J-7 (Mig-21 F) Fishbed C	26	36		
• Fighter Ground Attack (FGA)		296	138	108	200
	• JH-7	18	84		
	• JH-7/ JH-7A			84	120
	• Q-5 Fantan	30	30		
	• J-10A/S				28
	• J-11B/BS				24
	• Su-30 Mk-2 Flanker	48	24	24	24
	• J-6 (Mig-19S) FarmerB	200	-		
• Anti Tank (ATK)	Q-5 Fantan			30	
• Anti-submarine Warfare (ASW)	PS-5 (SH-5)	4	4	4	4
• Recce		7	13		
	• HZ-5 (Il-28R) Beagle	7	7	7	
	• Y-8J/Y-8JB Cub/ High New2	-	6	6	
• ELINT					7
	• Y-8J/Y-8JB Cub/ High New2				4
	• Y-8X				3
• AEW&C					6

	• Y-8J				4
	• Y8W New High 5				2
• Maritime Patrol (MP)	Y-8X	4	4	4	
• Tanker (TKR)					3
	• HY-6	3	3		
	• H-6 DU			3	3
• Transport		66	66	66	66
	• Y-8 (An-12 BP) Cub A (Medium)	4	4	4	4
	• Y-5 (An-2) Colt (Light)	50	50	50	50
	• Y-7 (An-24) Coke (Light)	4	4	4	4
	• Y-7H (An-26) Curl (Light)	6	6	6	6
	• Yak-42 (Light)	2	2	2	2
• Training		73	122	94	106+
	• JJ-6 (Mig-19 UTI) Farmer	16	14	14	14
	• CJ-6			38	38
	• HJ-5			5	5
	• JJ-7 Mongol A	4	4	4	4
	• PT-6 (CJ-6)	53	38		
	• HY-7		21	21	21
	• JL-8			12	12
	• JL-9				12+
	• K-8		12		
	• HJ-6		33		
Helicopters					
• Anti-Submarine Warfare/ Anti Submarine Unit Warfare (ASW/ ASUW)					44

	• Z-9C (AS-565SA) Panther	25	25	25	25
	• Ka-28 Helix	10	10	13	19
• AEW					10+
	• Ka-31				9
	• Z-8 AEW				1+
• Search & Rescue (SAR)/ Transport (TPT)		35	35	48	6
	• SA-321 (heavy)	15	15	15	15
	• Z-8, Z-8A (SA-321Ja) Super Frelon (heavy)	20	20	20	20
	• Z-8JH			3	4 (SAR)
	• Z-8S (SAR)			2	2
• Support	Mi-8 (Hip)	8	8	8	8
Marines		10,000	10,000	10,000	10,000
Forces by Role					
• Army	Divisions (also have amphibious role)	3	3	-	-
• Marine Infantry	Brigades (each; 1 infantry (inf) battalion (bn), 1 AD bn, 1 armed (mech) inf bn, 2 amph recce bn, 2 arty bn & 2 tank bn)	2	2	2	2
Equipment by Type					
• Light Tank	Type-63A	150	150		124
• APC(Tracked)	Type-63	60	60		248
• APC(Wheeled)	Type-92	nk	nk		
• Artillery Towed	Type-83 (122mm)	nk	nk		40+
• Multiple Rocket Launcher	Type-63 (107mm)	nk	nk		
• Anti Tank Missile	HJ-73 & HJ-8	nk	nk		
• AD SAM MANPAD	HN-5 Hong Nu/ Red Cherry	nk	nk		

Appendix 6

An Analysis of the Military Balance and Equipment Holdings

Overall Strength

	China	**India**
Active	22,85,000	13,25,000
• Army	16,00,000	11,29,900
• Navy	2,55,000	58,350
• Air Force	3,00,000 – 3,30,000	1,27,000
• Strategic Missile Force	1,00,000	
• Coast Guard		9,550
Para Military	6,60,000	13,22,150
Reserves	5,10,000	11,55,000 • Army – 9,60,000 • Navy – 55,000 • Air Force – 1,40,000 • Para Military – 9,87,000

Ground Forces

	China	India	Remarks
Command		Six Regional Command HQs & one Training Command HQ	
• Military Area Commands (MACs)	7		
• Combined Corps HQ	18		

• Strike Corps HQ		3	
• Holding Corps HQ		10	
Missile			
• Missile Gp with Agni I/ II		2	
• Missile Gp with SS – 150/ 250 Prithvi I/ II		2	
Special Forces	7 Units	8 SF Bn@	@in addition Special Frontier Force (strength 10,000) consisting of mainly ethnic Tibetans is also available.
Manoeuvre			
• Armoured			
• Armd Divs	5	3 (2-3 Armd Bdes & 1 SP Arty Bde (1 Med Regt & 1 SP Arty Regt))	
• Armd Bdes	11	8 (Indep)	
• Mechanised			
• Mech Inf Divs	7	4 RAPID (1 Armd Bde, 2 Mech Bde & 1 Arty Bde)	
• High Alt Mech Inf Div	2		
• Mech Inf Bdes	9	2 (Indep)	
• High Alt Mech Inf Bde	1		
• Indep Mech Inf Regt	2		
• Light			

• Inf Divs		17 (2-5 Inf Bdes & 1 Arty Bde)@	@ In addition following paramilitary forces are also available • Rashtriya Rifles – 5 Force HQs, 15 Sector HQs & 65 Battalions (under Ministry of Defence) • Assam Rifles – 7 Sector HQs & 42 Battalions (under Ministry of Home Affairs) • Border Security Force – 170 + Battalions (under Ministry of Home Affairs) • Indo Tibetan Border Police – 30 Battalions (under Ministry of Home Affairs)
• Mot Inf Div	10		
• High Alt Mot Inf Divs	3		
• Jungle Mot Inf Divs	1		
• Mot Inf Bdes	19		
• Inf Bdes		7	
• High Alt Mot Inf Bdes	2		
• Mountain			
• Mtn Divs		12 (3-4 Inf Bdes & 3-4 Arty Regts)	

• Mtn Inf Bde	2		
• Indep Mtn Bde		2	
• Air Manoeuvre		1 Para Bde	
• Amphibious			
• Amph Armd Div	1		
• Amph Mech Div	2		
• Other			
• OPFOR Armd Bde	1		
• Mech Gd Div	1		
• Lt Gd Div	1		
• Aviation			
• Avn Bde	6		
• Avn Regt	4		
• Trg Avn Regt	2		
• Hel Sqns		14	
Combat Support			
• Arty Div	2	3*	*2 Arty Bdes (3 Med Arty Regts & 1 STA/ MRL Regt)
• Arty Bde	17		
• Coastal Def AShM Regt	9		
• AD Bde	21	8	
• Engr Bde	1	4	
• Engr Regt	13		
• EW Regt	5		
• Sigs Regt	50		
• SSM Regt		2$	$with PJ-10 Brahmos Missiles

Summary

	China	India
Military Area Commands/ Command HQs	7	6 + I Training Command
Combined Corps (Group Armies)/ Corps	18	13
Infantry Divisions	14#	17
Infantry Brigades	21	7
Mountain Divisions		12
Mountain Brigades	2	2
Mechanized Infantry Divisions	9	4 (RAPID)
Mechanized Infantry Brigades	12	2
Armed Divisions	5	3
Armed Brigades	11	8
Artillery Divisions	2	3
Artillery Brigades	17	
Airborne Divisions	3	
Airborne Brigade		1
Amphibious Divisions	2	
Amphibious Brigades	1	

Analysis

- Force on force basis the advantage is with the PLA. In a conflict scenario, with a dormant Western front, there is a near parity between the two contestants. However in a collusive scenario the force ratio will be in China's favour since India will have to maintain a strategic balance on the Western front. A clear assessment should be done at the Service HQs to identify the force levels and the force multipliers that need to be deployed on the Pakistan front for strategic deterrence and lethal punitive options. The decision for timely mobilization of the dual task formations will be predicated on this assessment. The military balance 2013 shows that China has 5 Armd Divisions, 11 Armd Brigades, 9 Mechanized Infantry Divisions, 12 Mechanized Brigades, 14 Mot Infantry Divisions and

21 Mot Infantry Brigades. This is in contrast to the conventional assessment that China has approximately 45 to 46 divisions, out of which 32 to 33 divisions can be applied against India. Based on intelligence inputs available with Service HQs, another study should be undertaken to reconcile the figures viz – a – viz The Military Balance 2013.

- The PLA enjoys a comprehensive edge in terms of its higher level of mechanization and specialization of formations that can operate in high altitude terrain.

- The infrastructure and terrain allows the PLA to mobilize and deploy with a great ease and freedom.

- The air defence units with PLA and its reserves with an integrated air defence infrastructure provide the PLA with a more credible and effective air defence umbrella.

- The paramilitary and the militia in case of China is under command the CMC/ MAC. In our case the paramilitary is under command the Ministry of Home Affairs, which has its inherent problems of employment, deployment, training and command & control.

Strategic Forces

	China	India
Missiles	470	54
ICBM	72	In test
IRBM	2	24
MRBM	122	-
SRBM	252	30
LACM	54	Nirbhay
Warheads	250	90 - 110
Submarines		
SSBN	4	-
Satellites	**52**	**3**
Communications	4	-
Navigation/ Positioning/Timing	17	-
ISR	20	3
ELINT/ SIGINT	11	-

Analysis

- In strategic assets and space based capabilities there exists a wide gap to India's disadvantage.

- The ABM, ASAT and Anti Ship (DF 31D type) capability with Indian forces is in a nascent stage. The Brahmos is the only credible weapon system in this domain that too has a limited range (< 300 kms).

Analysis of Mechanized Forces

Equipment by Type	China	India
MBT	7,430+	3,274+
LT TK	800	-
AIFV	2,150	1,455+
APC	2,900	336+
APC (T)	2,000	-
APC (W)	900	157+
PPV	-	179
RECCE	-	-

An Analysis of Armour

China		India		Remarks
MBT	7,430+	**MBT**	3,274+	1,100 various models in store
• Type 59/ Type 59 D/ Type 59 II	4300			
• Type 79	300			
• Type 88 A/ B	500			
• Type 96	1000			
• Type 96 A	800			
• Type 98 A/ Type 99	500			
• Type 99 A2	30+			

		• Arjun	124	
		• T - 55	715	Being phased out
		• T – 72 M1	1,950	
		• T – 90S	485+	

- Type 59 is a 1950s vintage tank and is a Chinese produced version of the Soviet T – 54A tank, the earliest model of the T -54/ T – 55 series.

- Type – 59 II (WZ-120B). It has a 105 mm Type 81 rifled gun, a new radio and fire suppression system. Produced from 1982 to 1985.

- Type – 59D (WZ-120C). It was developed in the 1990s as an upgrade of existing T – 59s, fitted with explosive reactive armour, a new tank gun, passive night vision, new fire control and a 580 hp 12150L7 engine. Variants include Type 59D and Type 59D1.

- Type 79. Development of the T – 59 tank. It was the first independently-developed main battle tank by China. The improvements included a new engine, ballistic computer and laser rangefinder. The more advanced Type 79 was equipped with a 105 mm rifled gun.

- Type – 88. It is an improved version of the Type – 80 MBT, which was based in the Type – 79 tank. It entered service in 1988 and this tank is unique in that unlike the rest series of Chinese tanks, this series actually includes versions from different families of earlier tanks. Production of Type 88-series MBTs was stopped in 1995. In test its armour was 200 mm vs APFSDS and 300 mm vs HEAT. It has an autoloader which gives it a maximum rate of fire of 2.5 rounds a minute.

- Type – 96. This is a Chinese Third Generation main battle tank . Based on the Type 85-III design, the Type 96 entered service with the People's Liberation Army (PLA) in 1997.

- Type – 98 and Type – 99. These are the latest of the third generation tanks with the PLA and are based on T – 72/ T – 80 designs.

From the above it is evident that as on date there is no major asymmetry between the Indian Army and the PLA as far as the numbers of tanks is concerned. However from the employment point of view following merit attention:-

- The major part of our armoured formations is deployed on the Western borders. On our Northern borders with China space for deployment, inadequacy of induction routes and winterization of tanks (pre – heaters) are major constrains.

- The PLA is in an advantageous position to employ armour against us. The terrain and the available infrastructure allow the PLA to mobilize, build up and deploy its mechanized forces with ease in comparison to us.

Analysis of Artillery, Aviation and Air Defence Artillery including Missiles

Arty	12,367+	9,682+
SP	1,821	20+
Towed	6,140	2,970+
Gun/ Mortars	**50+**	**6,520+**
MRL	1,770+	192
SP	1,716	
Towed	54	6,520+
Mortor	2,586	
Aircrafts: Transport	8	
• Medium	3	
• Light	5	
Helicopters		
• ATK	42	-
• MRH	401	232
• **Transport**	**234**	-
▪ Heavy	21	-
▪ Medium	145	-

▪　Light	68	-
Air Defence (AD)		
SAM	302+	3,300+
SP	302	680+
MANPAD	-	2,620+
Guns　. Towed	7,700+	2,395+
Radars: Land	Cheetah	38+

An Analysis of Artillery

	China		India
Total	**12,367**		**9,682**
SP	**1,821**	SP	**20+**
122mm	1,371		
• Type – 70-1/ Type – 89/ Type – 07 (PLZ-07)	1,296		
• Type – 09 (PLC – 09)	75		
		130mm M – 46 Catapult	20+
152mm Type - 83	324	152mm 2S19 Farm	Reported
155mm Type – 05 (PLZ – 05)	126		
Towed	**6,140**	**Towed**	**2,970+**
		105mm	1,350
		IFG Mk 1/ Mk 2/ Mk 3	600+ (Being replaced)
		LFG	700
		M – 56	50

122mm Type – 54-1 (M – 1938/ Type – 83/ Type – 60 (D – 74)/ Type – 96	3,800	122mm D – 30	520
130mm Type – 59(M – 46)/ Type - 96	234	130mm M – 46	600 (500 in store)
152mm Type – 54(D – 1)/ Type – 66(D – 20)	2,106		
		155mm	500
		• FH – 77B	300
		• M – 46(modified)	200
Gun/ Mor	**50+**		
120mm Type – 05(PLL – 05)	50+		
MRL	**1,770+**		
SP	**1,716**	**SP**	**192**
• 107mm some 122mm Type – 81/ Type – 89/	1,620	122mm BM – 21/ LRAR	150
• 300 mm Type – 03 (PHL – 03)	96	214mm Pinaka	14 (non operational)
		300mm 9A52 Smerch	28
Towed			
107mm Type 63	54		
Mors	**2,586**	**Mors**	**6,520**

From the number of weapon systems it is clear that there is a major asymmetry between the Indian Army and the PLA. However the following merit attention:-

- There is a total asymmetry as far as the SP artillery is concerned. However in an India – China scenario this makes very little difference as its deployment and employment options are limited.

- As far as the field artillery and the MRLs are concerned the PLa has 122mm as its basic weapon system and Indian Army has 105mm, 130mm and 155mm as their basic weapon system. In Indian Army the 105 mm is in the process of being phased out and the 155mm will be the standard weapon system.

Naval Forces

Navy	China	India
	255,000	58,350
Submarines	65	-
Strategic: SSBN	4	-
Tactical	61	15
SSN	5	1
SSK	55	14
SSB	1	-
Principal Surface Combatants	77	21
Aircraft Carriers. CV	1	1
Destroyers	14	11
DDGHM	12	6
DDGM	2	5
Frigates	62	12
FFGHM	29	11
FFGM	1	-
FFGM	4	-
FFG	28	-
FFH	-	1
Amphibious	-	-
Principal Amphibious Vessels: LPD	2	1
Landing Ships	85	10
LSM	59	5
LST	26	5
Landing Craft	151	8

LCU	120	8
LCM	20	-
LCAC	1	-
UCAC	10	-
Logistics and Support Ships	**205**	**50**
Naval Aviation	**26,000**	**7,000**
Air Craft (combat capable)	**341**	**34**
BBR	30	-
FTR	72	15
FGA	200	10
ASW	4	9
ELINT	7	-
AEW&C	6	-
ISR	7	-
TKR	3	-
TPT	66	37
TRG	106+	12
MP	-	14
Helicopters		
ASW	44	54
AEW	10	9
SAR	6	-
TPT	43	-
MRH	-	53
Marines	**10,000**	**1,200**
LT TK	124	-
APC (T)	248	-
ARTY	40+	-

Analysis of Naval Forces

- There exists a vast asymmetry in terms of number of principal combatants and submarines (both nuclear and conventional) held.

- The Indian Navy has experience in operating an Aircraft Carrier, which the Chinese Navy lacks.

- PLAN has a number of nuclear submarines (both SSBN and SSN) and has modern conventional submarines with AIP (advanced Kilo class). Indian Navy has one Charlie class Russian nuclear submarine on lease, a non operational indigenous nuclear submarine and no conventional submarine with AIP.

- PLAN with PLASAF has developed a credible and effective area denial capability in the Yellow, East China and South China Sea's.

- PLAN is well equipped for littoral warfare and has an AD cover well beyond its maritime boundaries.

Air Force

	China	India
Total Strength	**300,000- 330,000**	**127,000**
Air Craft (combat capable)	**1,903**	**870**
FTR	842	63
FGA	543	736
ATK	120	-
EW	13	-
ELINT	4	-
ISR	51	3
AEW&C	8	3
C2	5	-
TKR (AAR)	10	6
TPT	326+	238
TRG	950	241

Helicopters

	China	India
MRH	22	226
TPT	28	106
ATK	-	20

Air Defence

SAM	600+	-
SP	300+	-
Towed	300+	-
Guns	16,000	-

Analysis of Air force

- The numbers do not portray the exact picture. Though the numbers suggest a wide asymmetry in the combat aircrafts, but the quality of combat and EW & C aircrafts with IAF certainly tilts the balance. This balance will be further tilt in India's favour with the induction of MMRCA into IAF.

- The altitude restrictions of operating from high altitude air fields forces the PLAAF to either operate with lower pay loads or deploy a number of AAR aircrafts to compensate for payload loss. This may also compel the PLA to increasingly base their air campaign on LACMs/ conventional SSMs. However this has a inherent danger of accidental triggering of a nuclear exchange.

Annexure 1

Revolution in Military Affairs with Chinese Characteristics

The PLA, aiming at building an informationalized force and winning an informationalized war, deepens its reform, dedicates itself to innovation, improves its quality and actively pushes forward the RMA with Chinese characteristics with informationalization at the core leading to changes in both structural and operational readiness. These are enumerated in the succeeding paragraphs.

Reducing the PLA by 200,000

It has been the established policy to build a streamlined military with Chinese characteristics. Since the mid-1980s, China has twice downsized its military by a total of 1.5 million. In September 2003, the Chinese government decided to further reduce 200,000 troops by the end of 2005 to maintain the size of the PLA at 2.3 million. The current restructuring, while cutting down the numbers, aims at optimal force structures, smoother internal relations and better quality.

The PLA will achieve rebalancing in the ratio between officers and men by streamlining structure, reducing the number of officers in deputy positions, filling officers' posts with non-commissioned officers (NCOs) and adopting a system of civilian employees, the number of the PLA officers can be substantially reduced to optimize the ratio between officers and men.

To improve the system of leadership and command the emphasis is put on streamlining the staff offices and the directly affiliated organs at the corps level and above, so as to compress the command chains and further improve the operational command system to strengthen the command functions. The numbers of offices and personnel are both reduced by about 15 per cent by adjusting staff functions, dismantling and merging offices and reducing the numbers of subordinate offices and assigned personnel.

To optimize the composition of the services and arms of the PLA, the Army is streamlined by reducing the ordinary troops that are technologically backward while the Navy, Air Force and Second Artillery Force are strengthened. The make-up of troops and the size of the services and arms are optimized with an increasing proportion of new- and high-tech units.

The PLA continues to adopt the system of joint logistical support at military area commands. The scope of joint logistical support is further enlarged and the number of logistical organizations and personnel are reduced while the rear hospitals, recuperation centers and general-purpose warehouses formerly under the administration of the services and arms are all integrated and reorganized into the joint logistical support system. An integrated tri-service joint logistical support system gradually takes shape, thus improving the overall efficiency.

The PLA aims at improving the structure and system for educating military personnel in both military and civilian educational institutions, and speeding up the establishment and improvement of a new educational system. This new system focuses on pre-assignment education which is separated from education for academic credentials. In accordance with the requirements for running educational institutions intensively on a proper scale, the PLA has optimized the system and structure of educational institutions by cutting down on those that are not essentially different from their civilian counterparts, and those that are more than necessary, and merging those that are co-located or have similar tasks.

Strengthening the Navy, Air Force and Second Artillery Force

While continuing to attach importance to the building of the Army, the PLA gives priority to the building of the Navy, Air Force and Second Artillery Force to seek balanced development of the combat force structure, in order to strengthen the capabilities for winning both command of the sea and command of the air, and conducting strategic counter-strikes.

The PLA Navy is responsible for safeguarding China's maritime security and maintaining the sovereignty of its territorial seas along with its maritime rights and interests. The Navy has expanded the space and extended the depth for offshore defensive operations. Preparation for maritime battlefield is intensified and improved while the integrated combat capabilities are enhanced in conducting offshore campaigns, and the capability of nuclear counter-attacks is also enhanced. In accordance with the principle of smaller but more efficient troops, the PLA Navy compresses the chain of command and reorganizes the combat forces in a more scientific way while giving prominence to the building of maritime combat forces, especially amphibious combat forces. It also speeds up the process of updating its weaponry and equipment with priority given to the development of new combat ships as well as various kinds of special-purpose aircraft and relevant equipment. At the same time, the weaponry is increasingly informationalized and long-range precision strike capability raised. It takes part in joint exercises to enhance its joint operational capabilities and integrated maritime support capabilities.

The PLA Air Force is responsible for safeguarding China's airspace security and maintaining a stable air defence posture nationwide. In order to meet the requirements of informationalized air operations, the Air Force has gradually shifted from one of territorial air defence to one of both offensive and defensive operations. Emphasis is placed on the development of new fighters, air defence and anti-missile weapons, means of information operations and Air Force automated command systems. The training of inter-disciplinary personnel is being accelerated for informationalized air operations. Combined arms and multi-type aircraft combat training is intensified to improve the capabilities in operations like air strikes, air defence, information counter-measures, early warning and reconnaissance, strategic mobility and integrated support. Efforts are being made to build a defensive air force, which is appropriate in size, sound in organization and structure and advanced in weaponry and equipment, and which possesses integrated systems and a complete array of information support and operational means.

The PLA Second Artillery Force is a major strategic force for protecting China's security. It is responsible for deterring the enemy from using nuclear weapons against China, and carrying out nuclear counter-attacks and precision strikes with conventional missiles. By upgrading missiles, stepping up the R&D of missiles, and promoting the informationalization of missiles and supporting equipment for command, communications and reconnaissance, the Second Artillery Force has built in its initial form a weaponry and equipment system that comprises both nuclear and conventional missiles, covers different ranges, and possesses markedly increased power and efficiency. The PLA Second Artillery Force boasts a contingent of talents mainly composed of academicians of the Chinese Academy of Engineering and missile specialists. More than 70 per cent of its active-duty officers have bachelor's degrees or above. High-tech means are used to reform its training and shorten the cycle for new weaponry and equipment to be combat-ready. It conducts missile-launching training and readiness exercises in near-real conditions and constantly enhances its quick-response and precision-strike capabilities.

Speeding Up Informationalization

In its modernization drive, the PLA takes informationalization as its orientation and strategic focus. By adopting the general approach of giving priority to real needs, making practical innovations, valuing talented personnel, and achieving informationalization by leaps and bounds, the PLA is actively engaged in the research and practice of informationalization.

In the past two decades, the PLA has been pushing forward informationalization in the field of military operations, focusing on command automation. It has completed a series of key projects to build military information systems and made great progress in building information infrastructure. As a result, command means have been substantially improved at all levels of headquarters and combat troops. Computers and other IT equipment have been gradually introduced into routine operations. The ability to provide operational information support has been greatly enhanced while more and more IT elements have been incorporated into main battle weapon systems. The CMC has approved and promulgated the Guidelines for the Development of Automated Command Systems of the Chinese People's Liberation Army and the Regulations of the Chinese People's Liberation Army on Automated Command Systems, defining the goals and relevant policies and statutes for developing automated command systems.

In the new stage of the 21st century, the PLA strives to comprehensively push forward informationalization with military information systems and informationalized main battle weapon systems as the mainstay and with military informationalization infrastructure development supported and guaranteed. In its drive for informationalization, the PLA adheres to the criterion of combat efficiency and the direction of an integrated development, enhances centralized leadership and overall planning, develops new military theories and operational theories while optimizing management system and force structure, updating systems of statutes and standards, and emphasizing training for informationalization. The PLA strengthens the building of military information systems and speeds up the informationalization of main battle weapon systems. It also makes full use of various information resources and focuses on increasing system interoperability and information-sharing capability. The PLA takes advantage of progress in government and social sectors in the field of informationalization, and establishes a scientific research and production system and information mobilization mechanism that integrates military and civilian efforts to promote the informationalization process of both the PLA and the government.

Accelerating the Modernization of Weaponry and Equipment

The PLA regards weaponry and equipment as the crucial material and technological basis for pushing forward the RMA with Chinese characteristics. In accordance with the national security needs, the PLA accelerates the modernization of weaponry and equipment, depending on national economic development and technological advance.

In order to strengthen the capability to win local wars under informationalized conditions, the PLA, in its development of weaponry and equipment, stresses the importance of capstone design, persists in taking informationalization as the leading force while advancing mechanization and informationalization simultaneously, and strives to build a streamlined, efficient and optimized modern weaponry system appropriate in size and optimal in structure.

Giving priority to the development of new- and high-tech weaponry and equipment. The PLA intensifies its R&D efforts and strengthens its innovative capability through self-reliance. It accelerates the R&D of new informationalized combat platforms and precision munitions, as well as electronic counter-measures equipment, and puts more effort into elevating the capabilities for precision strikes and information operations.

Accelerating the modification of old and outmoded weaponry. A number of old and outmoded weapons and equipment, which are backward in technology, poor in performance and no longer cost-effective in maintenance, are being phased out, and part of the active-service main battle weaponry is reconfigured on a selective, priority and phasal basis. By embedding advanced technology, developing new munitions, and integrating command and control systems, the PLA has restored or upgraded the tactical and technical performance of some current main battle weapons.

Continuously elevating integrated support for weaponry and

equipment. Taking existing weaponry and equipment as the basis, the PLA emphasizes the organic and systematic development of combat and support capabilities of weaponry and equipment. In accordance with the development of main battle weaponry and equipment, the PLA develops new types of general- and special-purpose support equipment, while strengthening the maintenance and technical support forces with priority given to new equipment and the training of personnel who employ, maintain and manage the new equipment, so as to elevate the integrated support of weaponry and equipment, thus satisfying the needs of readiness for military struggle.

Implementing the Strategic Project for Talented People

In August 2003, the CMC began to implement its Strategic Project for Talented People. The Project proposes that in one to two decades, the PLA will possess a contingent of command officers capable of directing informationalized wars and of building informationalized armed forces, a contingent of staff officers proficient in planning armed forces building and military operations, a contingent of scientists capable of planning and organizing the innovative development of weaponry and equipment and the exploration of key technologies, a contingent of technical specialists with thorough knowledge of new- and high-tech weaponry performance, and a contingent of NCOs with expertise in using weapons and equipment at hand. The Project will be implemented in two stages. By the end of 2010, there will be a remarkable improvement in the quality of military personnel, and a big increase in the number of well-educated personnel in combat units. The following decade will witness a big leap in the training of military personnel.

In recent years, the PLA has utilized military educational institutions as major platforms for training military personnel. Officer candidates have, in the main, been trained in four-year colleges. A functional transformation of military educational institutions is taking place with the emphasis shifting from academic credentials education to pre-assignment training. More and more military personnel with specialties for both military and civilian use will be trained by regular institutions of higher learning. So far, more than 90 such institutions have undertaken the task of training PLA cadres. In implementing the Project for Strengthening the Military with High-Caliber Personnel, nearly 30 key regular institutions of higher learning have trained a great number of Master Degree students for the

PLA, whose specialties are urgently needed. Various training courses have been offered at military educational institutions, including courses for young and middle-aged cadres, high-tech knowledge training courses for leading cadres at the levels of military area command and corps, and training programs of cross-service and cross-arm expertise. Hundreds of military cadres have been sent to the central and provincial Party schools. Division and brigade commanding officers have been arranged for study tours abroad. The number of commanders has been increased among the overseas military students.

Intensifying Joint Training

Adapting to the features and patterns of modern warfare, the PLA has intensified joint training among services and arms at all levels to enhance joint fighting capabilities.

Highlighting joint operational training. In view of the future operational tasks, the PLA has given priority to training with specific objectives, joint operational training and high-level command post training. It has successfully organized a series of major joint operational training activities. Studies and exercises directed at operational issues are emphasized with additional attention to the development of operational doctrines and training regulations, and the construction of network systems. By exploring approaches for operational guidance, operational command and operational training for joint campaigns, the PLA has improved the capabilities of commanding officers at each level to organize and direct joint operations.

Conducting joint tactical training. To meet the needs of joint operations at the tactical level, units of different arms and services stationed in the same areas have intensified their contacts and cooperation in the form of regional cooperation to conduct joint tactical training. In September 2003, the General Staff Headquarters organized a PLA-wide demonstration on regional cooperation for military training in Dalian. That event drew lessons from regional cooperation for military training and explored new ways to conduct joint tactical training.

Improving the means of joint training. After years of development, substantial progress has been achieved in on-base training, simulation training and network training. Almost all combined tactical training activities at division, brigade and regiment levels can be conducted on base.

All services and arms have set up their basic simulation training systems for operational and tactical command. A (joint) combat laboratory system of simulation training for all military educational institutions has been initially put in place. A military training network system has been set up to interconnect the LANs of military area commands, services and arms, and command colleges.

Training commanding officers for joint operations. The military educational institutions have intensified their joint operations training. The elementary command colleges offer basic courses in joint operations. The intermediate command colleges offer courses on service campaigns and combined operations. The advanced command university offers courses on strategic studies and joint operations. In order to bring up commanding officers for joint operations, PLA units carry out on-duty training and regional cooperation training, and acquire knowledge of other services and arms and joint operations through assembly training, cross-observation of training activities, academic seminars and joint exercises.

Deepening Logistical Reforms

The PLA continues to deepen, expand and coordinate the reforms of its logistical system, and makes efforts to enhance the capability to provide fast, efficient and integrated support.

Pushing forward an integrated tri-service support system. Experimental reforms of joint logistics started in the Jinan Theater in July 2004. First, all logistical organs of the three services are integrated into one. The Theater Joint Logistics Department or Joint Logistics Department of Military Area Command, originally called Logistics Department of Military Area Command, takes responsibility for joint logistical support for all in-theater units of the three services. The percentage of non-Army cadres in this department has risen from 12 per cent to 45 per cent. Second, all logistical support resources of the three services are integrated. All in-theater logistical support facilities such as rear depots, hospitals, recuperation centers, and material supply and engineering facilities, originally under the leadership and management of the services and arms, have been transferred to the joint logistics system for unified integration, construction, management and employment. Third, all logistical support mechanisms of the three services are integrated. The in-theater logistical support for troops of the three services is no longer categorized into general or special supply support. All supplies are planned and provided by the joint logistics system. Fourth, all logistical support channels of the three services are integrated. The multiple support channels for troops of the arms and services have been readjusted and integrated into one support channel of the joint logistics system, aiming at compressing the supply chain and improving efficiency to form an effective system of supervision and management.

Conducting technological research of logistical equipment. Over the past two years, the PLA has completed experiments to finalize the designs of 92 types of new logistical equipment, with the designs in logistical equipment system finalized at a rate of 93 per cent. A new-generation logistical equipment system with all necessary specialized varieties has been basically established with some of the equipment reaching the internationally advanced standards. The Fourth Beijing International Exhibition on Military Logistical Equipment and Technology was held in April 2004. More than 340 manufacturers from 26 countries and regions took part in the exhibition, and military logistics delegations from 16 countries were invited to attend the exhibition as well as the international symposium on the development strategy of military logistical equipment and technology.

Deepening reforms of the medical support system and logistics outsourcing. In May 2004, the PLA started in an all-round way to carry out the reform of its medical support system based on pilot and expanded experiments. The reform features categorized support, appropriate medical care, unified management and treatment at designated hospitals, and treatment upon presentation of medical cards. The PLA has established a new type of medical support system in which medical service is free for servicemen, preferential for dependents accompanying officers, and available to civilian employees in the PLA through medical insurance. This has improved the quality of medical service and enhanced the capability of medical support. The PLA has adopted the management method of packaging wages for civilian employees and the policy of providing resettlement benefits to redundant personnel. It has also introduced in an all-round way such housing reform measures as monetization, market supply and management outsourcing, stepped up efforts to cash housing subsidies, and further enabled servicemen to purchase houses.

Innovating Political Work

The PLA takes as guidance Marxism-Leninism, Mao Zedong

Thought, Deng Xiaoping Theory and the important thought of the "Three Represents," adheres to the fundamental principle and system of the Party's absolute leadership over the armed forces, puts ideological and political work first, innovates political work in its content, approaches, means as well as mechanism to give full play to the support and combat functions of political work.

In December 2003, the new Regulations on the Political Work of the Chinese People's Liberation Army was revised and promulgated. The regulation maintains that political work is the fundamental guarantee of the Party's absolute leadership over the armed forces and the assurance for the armed forces to accomplish their missions. It clearly defines political work as a significant component of combat capabilities of the PLA, and stresses the importance of giving full play to the combat function of political work. Education in the RMA with Chinese characteristics is given PLA-wide. Wartime political work is studied and rehearsed extensively. Political work is strengthened in all services and arms as well as the units carrying out special missions. Education in the PLA's functions and sense of urgency has been intensified in the PLA so that officers and men are motivated in their trainings and a tough fighting spirit and a good working style are fostered.

The PLA relies on laws and regulations to promote the innovation of political work. In April 2004, the CMC promulgated the Regulations on the Work of the Armed Forces Committees of the Communist Party of China (for Trial Implementation), which further defines the duties and responsibilities of the Party committees, the standing committees of the Party committees, secretaries and committee members, and further improves the decision-making procedures and principles in Party committees. In February 2004, the CMC released the Provisions on Strengthening the Education and Management of High- and Middle-Ranking Cadres of the PLA, which establishes and refines the systems for cadres at the regiment level and above to do self-study and review, to receive thematic education, to take admonishment talks, to make ideological and political assessment, to submit work and probity reports as well as reports on important work assignments.

The PLA attaches great importance to ideological and cultural work. In the period of 2000-2002, the CMC allocated RMB 140 million for the cultural work of grass-roots units. In recent two years, the General Political

Department and the General Logistics Department have jointly issued a number of regulations in succession, including the Provisional Regulations of the Chinese People's Liberation Army on the Management of Cultural Equipment and the Provisional Regulations on Grass-Roots Cultural Construction. Beginning in 2003, the cultural equipment supplied to grass-roots units are covered by regularized outlays and managed as organic equipment. In May 2004 a PLA-wide forum on art and literature was held, in which a five-year plan was formulated for art and literature work in the military. The PLA publishes more than 2,800 titles of books and audio-visual products every year. All units carry out rich and colorful on-camp cultural activities to promote the all-round development and enhance combat capability.

Governing the Armed Forces Strictly and According to Law

The PLA implements the principle of governing the armed forces strictly and according to law, strengthens the building of the military legal system, raises the level of regularization, and enhances the combat capability of the armed forces.

The PLA has emphasized incorporating into laws and regulations its good traditions in governing the armed forces and the requirements of the RMA with Chinese characteristics, so as to regulate all dimensions of the armed forces building. In the new historical era, the PLA has promulgated and revised a large number of military regulations, including the Regulations on Routine Service of the People's Liberation Army, Regulations on Discipline of the People's Liberation Army, Regulations on Formation of the People's Liberation Army, Regulations on the Headquarters of the People's Liberation Army, Regulations on the Political Work of the People's Liberation Army, Regulations on the Logistics of the People's Liberation Army, Regulations on the Armaments of the People's Liberation Army, Regulations on the Military Training of the People's Liberation Army, Regulations on the Garrison Service of the People's Liberation Army, and a new generation of operations regulations. The military law system has been basically established with regulations as its main body. In April 2003, the CMC promulgated the Regulations on Military Rules and Regulations to regulate the military legislative work. In January 2004, according to the CMC's directive, the PLA and the People's Armed Police Forces (PAPF) proceeded to sort out in a comprehensive way all their current regulations and rules, and uniformly organize the compilation and printing of the collections of military regulations and rules so as to provide legal basis for strict governing of the armed forces. The armed forces have carried out legal education in a deep-going way and conducted regulation training courses at various levels to guide the officers and men to perform their duties in accordance with the law.

The PLA has maintained the authority and solemnity of the regulations and rules and administered troops strictly in accordance with the regulations and rules. Incorporating the cultivation of good style and strict discipline into routine military training and administration has helped to sharpen the awareness of the officers and men in their observance of regulations and rules. Through strict training, refined military bearing, strict discipline and resolute and swift work style have been cultivated among the troops. In August 2003, the CMC revised and issued the Outline for Armed Forces Building at the Grass-Roots Level, which has promoted the regularization of the orders in preparation against war, training, routine work and everyday life at the grass-roots level. The General Staff Headquarters, the General Political Department, the General Logistics Department and the General Armaments Department have twice formed joint working groups for overall inspection of strict administration of the troops. In accordance with the CMC requirements, the PLA and the PAPF have intensified rectification and improvement, and have further promoted the implementation of the guiding principle of governing the armed forces strictly and according to law.

Bibliography

Books

1. When China rules the World by Micheal Jacqus

2. Next 100 Years by George Friedman

3. China's Military Modernization: Building for Regional and Global Reach by Richard D Fisher Jr

4. Modernizing China's Military by David Shambaugh

5. Modernization of the PLA by JS Bajwa

6. When China meets India by Thant Myint U

7. Chinese Views of Future Warfare by Micheal Pillsbury

8. Modern Chinese Warfare 1785 – 1989 by Bruce A Elleman

9. The Science of Military Strategy by Peng Guanrqian and Yao Youzhi

10. Unrestricted Warfare by Qiao Liang and Wang Xiangsui

11. The Chinese Army Today: Tradition and Transformation for the 21st Century by Dennis J Blasko 2006 and Revised edition 2013

12. PLA as Org : Ref Vol Ver 1.0 by Mulvenon Jameric and Yang Andrew

13. Rising to the Challenge: China's Grand Strategy and International Security by Avery Goldstein

14. The Grand Strategy of the Byzantine Empire by Edward N Luttwak

15. Worse than a Monolith: Alliance Politics and Problems of Coercive Diplomacy in Asia by Thomas J. Christensen

16. The Grand masters Insight on China and the US by Lee Kuan Yew.

17. Chinese and Indian Strategic Behavior by George J. Gilboy & Eric Heginbotham.

18. Chinese Warfighting: The PLA Experience since 1949 by Mark A Rayn, David M Finkelstein & Michael A McDevitt.

19. The Dragon extends its Reach by Larry M. Wortzel

20. China Goes Global: The Partial Power by David Shambaugh

21. Ancient Chinese Thought, Modern Chinese Power by Yan Xuetong

22. Sun Tzu and Art of Modern Warfare by Ralph D Sawyer

23. Interpreting China's Military Power; Doctrine Makes Readiness by Ka Po Ng

24. The Seven Military Classics of Ancient China Ralph D Sawyer

25. The Military Balance by International Institute of Strategic Studies, London from 2009 to 2014

Reports and White Papers

26. White Papers on China's National Defence 1998 to 2010.

27. The Diversified Employment of China's Armed Forces, Defence White Paper 2013.

28. Annual Reports to Congress: Military and Security Developments Involving the Peoples Republic of China 2008 to 2014.

29. Report to the Congress of the US-China Economic and Security Review Commissions 2008 to 2013.

30. Congressional Hearings of the Executive Commission on China of the US Government

Index

Symbols

15th Airborne (AB) Corps 80

A

Academy of Military Sciences xv

Air and Space Integrated Operations 84

Airborne early Warning xv

Airborne Early Warning and Control xv

Air to air refuelling xv

Aksai Chin 13, 136, 137, 139, 153

Anti-Tank Guided Missile xv

Autumn Harvest Uprising 5, 6

B

Border Control Department 64

Brezhnev Doctrine 14

Bronze Age xi

C

Central Military Commission xv, 6, 9, 10, 15, 16, 23, 24, 27, 28, 33, 61, 63, 64, 65, 68, 76, 150, 202, 215, 217, 221, 222, 223

Central Military Department 6

Chengdu MAC 55, 59, 61, 69, 148, 149, 159, 162

Chengdu Military Region 44, 45

China Railway High-speed 60

Chinese Communist Party 7, 28, 31, 134, 227

Chinese Communist Party rule 31

Chinese Military Satellites vii, 82

Chinese People's Volunteers xv

Command, Control, Communications and Intelligence xv

Communist Party of China xv, 2, 5, 6, 8, 9, 11, 33, 34, 82, 134, 135, 155, 221

Counter-terrorism 64, 70

Cultural Revolution 2, 13, 14, 20, 76, 94

D

David Shambaugh 39, 225, 226

Defensive military strategy 31

Deng Xiaoping ix, 10, 15, 18, 19, 21, 22, 37, 76, 100, 221

Dennis J Blasko 39, 72, 161, 165, 225, 227

Desert Storm 76

Dong Feng xvi, 120

Downsizing of the PLA 34

E

East China Sea Fleet 61

East Sea Fleet vii, 110, 171

Electronic countermeasures 78

Electronic Intelligence xvi

F

Force Mobility 43

Fujian Province 6

G

General Armaments Department xvi, 223

General Logistics Department xvi, 10, 222, 223

General Political Department xvi, 10, 15, 221, 223

General Staff Department xvi

Global force projection 31

Great Leap Forward 2

Group Army xvi, 45, 159

Guangzhou MAC 53, 59, 60, 148, 149, 151

Gulf War 25, 38, 229

Gutian Meeting 6

H

Hainan Island 10, 14

Hongqi-15 (HQ-15) 86

Huang he xi

Hu Jintao 31, 36

Hunan 5, 6, 182, 183, 184

I

Ilysuhin xvi

Immediate Action Unit 64

Information Warfare xvi

Intermediate Range Ballistic Missile xvi, 20, 121, 122, 126, 127, 129, 202

J

Jiangxi 5, 6, 7, 180, 183

Jiang Zemin 18, 27, 28

Jinan MAC 50, 59, 60, 148, 149

Jinan Military Region 45, 46

Jinggang Mountains 5

K

Korean War 11, 16, 19, 37, 76

Kuomintang xvi, 5, 6, 7, 8, 10

L

Lanzhou MAC v, 56, 59, 60, 135, 148, 149, 159, 162

M

Mao Zedong 5, 6, 7, 8, 12, 15, 18, 19, 37, 76, 155, 220

Medium Landing Ship xvi

Military Area Commands v, vii, xvii, 10, 14, 33, 34, 38, 39, 41, 43, 47, 49, 50, 51, 53, 55, 56, 58, 59, 60, 61, 69, 73, 108, 135, 148, 149,

150, 151, 159, 160, 162, 166, 197, 202, 213, 219

Military Districts xvii, 33, 48, 50, 51, 52, 54, 56, 57, 58, 116, 135, 160, 161, 183

Military Modernisation 11, 15, 39, 116

Military Regions 14, 24, 40, 44

Beijing 13, 14, 34, 36, 38, 45, 47, 48, 58, 59, 60, 65, 77, 81, 92, 94, 97, 98, 99, 103, 104, 105, 106, 108, 115, 116, 131, 135, 138, 140, 148, 149, 158, 167, 175, 177, 220

Chengdu v, 10, 14, 39, 44, 45, 55, 58, 59, 61, 69, 92, 93, 96, 135, 148, 149, 150, 159, 160, 161, 162, 164, 172, 184

Fuzhou 14, 166, 192

Guangzhou 6, 14, 34, 38, 53, 59, 60, 90, 92, 93, 111, 148, 149, 151, 171, 182, 184, 187, 192

Jinan 14, 38, 45, 46, 50, 58, 59, 60, 92, 93, 100, 108, 148, 149, 169, 179, 219

Kunming 14, 135, 159, 160, 161, 166, 185

Lanzhou v, 10, 14, 38, 56, 58, 59, 60, 69, 92, 93, 135, 148, 149, 150, 159, 160, 161, 162, 164, 168, 177, 179, 187

Nanjing 14, 34, 38, 45, 46, 51, 58, 61, 77, 92, 93, 110, 148, 149, 151, 170, 180, 182

Shenyang 14, 38, 49, 58, 59, 60, 77, 92, 96, 108, 148, 149, 166, 174

Wuhan 5, 14, 90, 166, 183

Xinjiang 10, 14, 57, 135, 136, 149,

160, 161, 162, 164, 178

Military Unit Code Designator xvii

N

Nanchang Uprising 5, 6, 7

Nanjing Military Area Command 61

Nanjing Military Region 45, 46, 110

National Defense Mobilization Committee xvii

National Defense University xvii, 24, 98, 118

Net Centric Warfare xvii

Ningxia Hui Autonomous Region 60

Northern Sichuan Province 7

North Sea Fleet vii, 108, 109, 167, 168

P

Paracel Islands 15

People's Armed Forces Departments xvii, 25

People's Armed Police Force v, xvii, 4, 38, 62, 63, 64, 65, 66, 67, 68, 69, 71, 162, 222, 223

People's Liberation Army ix, xvii, xviii, 3, 5, 9, 13, 34, 38, 61, 63, 67, 73, 76, 79, 89, 115, 116, 132, 161, 165, 204, 215, 221, 222

Air Force (PLAAF) Modernisation v, 75

Army (PLAA) Modernisation v, 37

Future Trends PLAN vii, 112

Historical Perspective v, 5

Navy (PLAN) Modernisation v, 100

Second Artillery Force (PLASAF) Modernisation v, 119

War Fighting Doctrines v, 5, 31

People's Militia 38

People's Republic of China xii, xviii, 10, 28, 36, 37, 38, 62, 63, 68, 72, 74, 78, 98, 108, 115, 116, 117, 120, 131, 134, 148, 155, 156, 226

People's War 9, 18, 20, 22, 37

Peoples War School 37

Peoples War under Modern Conditions 18, 20, 22, 26

Persian Gulf War 38

PLA v, vii, ix, x, xii, xvii, xviii, 2, 3, 4, 5, 6, 7, 8, 9, 10, 11, 12, 13, 14, 15, 16, 17, 18, 19, 20, 21, 22, 23, 24, 25, 26, 27, 28, 29, 30, 31, 32, 33, 34, 35, 36, 37, 38, 39, 41, 42, 43, 44, 45, 46, 54, 59, 60, 61, 62, 63, 64, 65, 66, 67, 68, 69, 70, 71, 72, 75, 76, 77, 78, 79, 85, 86, 87, 89, 94, 97, 98, 99, 100, 101, 102, 103, 104, 105, 106, 107, 108, 115, 119, 120, 121, 122, 124, 125, 126, 127, 132, 135, 137, 140, 150, 151, 157, 158, 161, 165, 201, 202, 204, 205, 207, 211, 212, 213, 214, 215, 216, 217, 218, 219, 220, 221, 222, 223, 225

PLA Air Force v, vi, vii, viii, ix, xvii, 3, 4, 33, 41, 43, 46, 60, 68, 75, 76, 77, 78, 79, 80, 81, 82, 83, 84, 85, 86, 87, 88, 89, 91, 92, 93, 94, 96, 97, 98, 99, 114, 152, 166, 174, 175, 178, 211, 214

PLA Army v, vii, ix, xvii, 3, 4, 33, 34, 37, 38, 39, 40, 41, 42, 44, 45, 46, 47, 58, 61, 62, 70, 71, 72, 100, 102, 148, 150, 158

PLA military Modernisation program 3

PLA Navy v, vi, vii, viii, ix, xviii, 3, 4, 21, 33, 34, 41, 61, 68, 84, 96, 100, 101, 102, 103, 104, 105, 106, 107, 108, 112, 113, 114, 115, 130, 151, 186, 210, 213

PLA Second Artillery Force v, ix, 3, 4, 114, 119, 120, 124, 126, 130, 131, 210, 214

PLA War fighting Doctrines vii, 18

Precision-guided munitions 30, 40, 77, 88

President Xi xi, 33

Q

Qinghai-Tibet railway 44

R

RAND Corp 79

Reserve Force 67, 68

Revolution in Military Affairs v, vi, xviii, 4, 5, 25, 26, 27, 29, 37, 38, 212, 216, 221, 222

S

Search and Rescue xviii

Second Artillery Corps 3

Shaanxi Province 8

Shangshu xi

Shannxi Province 9

Shenyang MAC 49, 59, 60, 148, 149

Shenyang Military District 77

Short-Range Ballistic Missile xviii

Snow Wolf Commando Unit 64

South China Sea ix, 2, 15, 61, 78, 106, 151, 210

South China Sea Fleet 61

South Sea Fleet vii, 111, 152, 172

Special Operations Forces xviii, 71

Special Police Units 64

Strategic Force Projection Capability 84

Surface-to-air missiles 40

    S-300 PMU-2, HQ-9, S-400 and the HQ-19 40

Surface to Surface Missile xviii

T

Taiping Rebellion 9

Taiwan 2, 46, 51, 58, 71, 77, 78, 93, 96, 100, 101, 106, 117, 149, 151

Taiwan Strait 77, 100, 106

Tank Landing Ship xvi

Taonan Tactical Training Base 60

'Tianxia' concept xi

Transformation of Group Armies vii, 43

Trans Region Support Operations 32

U

Unmanned Aerial Vehicle xviii, 40, 75, 97, 115

Unmanned Combat Aerial Vehicle xviii

W

Worker-Peasant's Red Army 7

Worker's and Peasant's Revolutionary Army 6

X

Xiangjiang River 60

Xi Jinping 33, 156

Y

Ye Jianying 6

Yellow river xi. *Yellow river* Huang he

Z

Zhang Tailei 6

Zhou Enlai 5, 15

Zhu De 5, 6

Zhuhai Air show 86

Zunyi Meeting 8